Disclaimer

The publisher of this book is by no way associated with the National Institute of Standards and Technology (NIST). The NIST did not publish this book. It was published by 50 page publications under the public domain license.

50 Page Publications.

Book Title: Report on Forum 2000: Fluid Properties for New Technologies - Connecting Virtual Design with Physical Reality

Book Author: James C. Rainwater; Daniel G. Friend; Howard J. Hanley; Allan H. Harvey; C D. Holcomb; Arno R. Laesecke; Joe W. Magee; Chris D. Muzny;

Book Abstract: Forum 2000, which addressed the present needs and priorities for thermophysical properties measurements, was held June 29, 2000, at the 14th Symposium for Thermophysical Properties in Boulder, Colorado. Seven distinguished experts presented overviews of issues related to a wide variety of subjects, and three discussion periods were held. This report documents in detail those overviews and discussion periods. Topics included whetehr simulation can replace experiment, properties needs for new miniaturization technologies, urgent problems such as nuclear waste cleanup, data needs for electrolyte systems and new generations of electric power plants, and data needs for unconventional materials such as molten metals and soft solids. Also included in this report are invited essays on issues raised at the Forum by panelists, audience participants, and other experts in thermodynamics.

Citation: NIST SP - 975

Keyword: electrolytes;Forum 2000;miniaturization;nuclear waste cleanup;power plants;simulation;soft solids;thermophysical properties

NIST Special Publication 975

Report on Forum 2000: Fluid Properties for New Technologies — Connecting Virtual Design with Physical Reality

James C. Rainwater
Daniel G. Friend
Howard J.M. Hanley
Allan H. Harvey
Cynthia D. Holcomb
Arno Laesecke
Joseph W. Magee
Chris Muzny

NIST CENTENNIAL
1901-2001

NIST National Institute of Standards and Technology • Technology Administration • U.S. Department of Commerce

NIST Special Publication 975

Report on Forum 2000: Fluid Properties for New Technologies — Connecting Virtual Design with Physical Reality

James C. Rainwater
Daniel G. Friend
Howard J.M. Hanley
Allan H. Harvey
Cynthia D. Holcomb
Arno Laesecke
Joseph W. Magee
Chris Muzny
Physical and Chemical Properties Division
Chemical Science and Technology Laboratory

September 2001

U.S. Department of Commerce
Donald L. Evans, Secretary

National Institute of Standards and Technology
Karen H. Brown, Acting Director

Certain commercial entities, equipment, or materials may be identified in this
document in order to describe an experimental procedure or concept adequately. Such
identification is not intended to imply recommendation or endorsement by the
National Institute of Standards and Technology, nor is it intended to imply that the
entities, materials, or equipment are necessarily the best available for the purpose.

National Institute of Standards and Technology Special Publication 975
Natl. Inst. Stand. Technol. Spec. Publ. 975, 122 pages (September 2001)
CODEN: NSPUE2

U.S. GOVERNMENT PRINTING OFFICE
WASHINGTON: 2001

For sale by the Superintendent of Documents, U.S. Government Printing Office
Internet: bookstore.gpo.gov   Phone: (202) 512-1800   Fax: (202) 512-2250
Mail: Stop SSOP, Washington, DC 20402-0001

Some figures in this publication are in color only in the electronic version on the enclosed CD-ROM or at

`http://Forum2000.Boulder.NIST.Gov/NISTSP975.pdf`

Foreword

As one of the organizers and chairs of the Fourteenth Symposium on Thermophysical Properties, I am pleased that Forum 2000 was an integral part of this Symposium. The Symposium of Thermophysical Properties has a long history and tradition as a focal point for the exchange of information on the thermophysical properties of fluids. One of the major outcomes of the Symposium has been to guide the future research activities in this area. Thus, the objectives of Forum 2000 meshed ideally with the goals of the Symposium.

As the Fourteenth Symposium on Thermophysical Properties was the last symposium of the twentieth century, two of its major objectives were (1) to take a retrospective view of the history of thermophysics during the last century and (2) to use this knowledge to make our best predictions about the future directions of thermophysics into the next century. This balance was achieved by several means, of which Forum 2000 was a key component.

The Keynote Speaker for the Fourteenth Symposium, Bill Wakeham of Imperial College, United Kingdom, set the tone for the conference. He was also a member of the panel for the Forum 2000 session. We are indebted to him and the other panel members for laying out a roadmap that will be both a challenge and an inspiration to our future research in fluid properties. The conclusions from the Keynote Address were expanded upon by those from Forum 2000. Wakeham emphasized the need to understand the theory and physics in the field of fluid properties, especially in the links between simulation or experiment and theory. Also, the challenges of new complex systems at extreme conditions should motivate new ideas and insights and provide an impetus for exciting new research directions into the next century.

The organizers of Forum 2000 have provided a valuable service to the thermophysical properties community in addressing the current needs and priorities for thermophysical properties research and in exploring the challenges facing this community with emerging new technologies. Many new areas have been identified for which fluid properties research can make significant contributions to satisfy technological needs. These include molecular simulation, metrology and combinatorial methods, process miniaturization, environmental technology, unconventional mixtures and materials, and data standardization and exchange.

The Symposium on Thermophysical Properties, as the major international conference on fluid properties, represents a primary mechanism for exchange of information among academic, industry, and government researchers and technologists. It is hoped that the results from Forum 2000 provide the stimulus and catalyst for the representatives of these communities working together to respond to the challenges in a timely manner to satisfy the new technological needs as we move into the next century.

W. M. Haynes
Chair and Organizer
Fourteenth Symposium on Thermophysical Properties
National Institute of Standards and Technology
Boulder, Colorado, USA

Forum 2000

Contents

Table 1: Panelists vii

Table 2: List of Audience Participants viii

1. Introduction 3
2. Molecular Modeling 5
    Peter Cummings
3. Microchemical and Thermal Systems: Process Intensification Through Miniaturization 17
    Ward TeGrotenhuis
4. First Discussion Period 23
5. Nuclear Waste Cleanup 29
    James Poppiti
6. Electrolytes 35
    Andrzej Anderko
7. Second Discussion Period 41
8. New Generation of Power Plants 45
    Thomas O'Brien
9. A Process Manufacturing Perspective 53
    Paul Mathias
10. Unconventional Materials 61
    William Wakeham
11. Forum Summary 67
    Howard Hanley
12. Third Discussion Period 69
13. Conclusions 75
14. Contributed Essays

    Thermophysical Property Needs for New Technologies: Electrolyte Systems 79
        Andrzej Anderko

    Comments on Forum 2000, Fluid Properties for New Technologies: Connecting Virtual Design with Physical Reality 82
        John R. Cunningham

    Dynamic Compilation: A Key Concept for Future Thermophysical Data Evaluation 83
        Michael Frenkel

An Academic Perspective on Experimental Thermophysical Properties Measurements 85
    James C. Holste

Thermophysical Properties – The Nature of the Science and Art 88
    Richard T Jacobsen

Micro, Nano, Pronto, Combi - What Thermophysicists (Can) Learn from Genomics 90
    Arno Laesecke

Improving the Physical Properties Infrastructure for Industry 93
    Alvin H. Larsen and George H. Thomson

Comments on the Status of Physical Properties for Chemical Manufacturing 97
    Paul Mathias

Simulations and Sensitivity Analysis 100
    Ray Mountain

The Importance of Experimental Measurements 102
    James D. Olson

Is the Job Complete? Definitely not! 105
    James C. Rainwater

Challenges in the Development of Transferable Force Fields for Phase Equilibrium Calculations 110
    J. Ilja Siepmann

A Perspective on Connecting Virtual Design with Physical Reality 113
    Lambert J. Van Poolen

Table 1: Panelists

Professor Peter T. Cummings (ptc@utk.edu)
Departments of Chemical Engineering, Chemistry, and Computer Science
University of Tennessee-Knoxville, USA
    and
Chemical Technology Division
Oak Ridge National Laboratory
Oak Ridge, Tennessee, USA

Dr. Ward TeGrotenhuis (Ward.Tegrotenhuis@pnl.gov)
Chief Engineer, Environmental Technology Division
Pacific Northwest National Laboratory
Richland, Washington, USA

James A. Poppiti (James.Poppiti@em.doe.gov)
Formerly Team Leader In-Tank Characterization
U. S. Department of Energy, Office of River Protection, Richland Operations Office
Hanford Site, Washington, USA

Dr. Andrzej Anderko (aanderko@olisystems.com)
Vice President, Properties of Fluids and Materials
OLI Systems, Inc.
Morris Plains, New Jersey, USA

Dr. Thomas J. O'Brien (Thomas.OBrien@netl.doe.gov)
Physical Scientist
U. S. Department of Energy, Office of Science and Technology
National Energy Technology Laboratory
Simulation & Multi-Phase Flow Analysis Division
Morgantown, West Virginia, USA

Dr. Paul M. Mathias (Paul.Mathias@aspentech.com)
Principal Advisor
Aspen Technology, Inc.
Cambridge, Massachusetts, USA

Professor William A. Wakeham (W.A.Wakeham@soton.ac.uk)
Formerly Pro Rector (Research)
Imperial College of Science, Technology, and Medicine
London, UK
Since October 1, 2001
Vice-Chancellor
University of Southhampton
Southhampton, UK

Table 2: List of Audience Participants

John Cunningham - Invensys Process Systems (formerly Simulation Sciences Inc.)

Dan Friend - NIST Boulder

Allan Harvey - NIST Boulder

James Holste - Texas A&M University

Richard Jacobsen - Idaho National Engineering and Environmental Laboratory

Kevin Joback - Molecular Knowledge Systems

Stephan Kabelac – Universität der Bundeswehr Hamburg (Germany)

John Kincaid - SUNY Stony Brook

Arno Laesecke - NIST Boulder

Ray Mountain - NIST Gaithersburg

Stephanie Outcalt - NIST Boulder

John Pellegrino – formerly NIST Boulder

Jim Rainwater - NIST Boulder

Stanley Sandler - University of Delaware

Ilja Siepmann - University of Minnesota

Truman Storvick - University of Missouri

George Thomson - DIPPR

Report on Forum 2000:
Fluid Properties for New Technologies —
Connecting Virtual Design with Physical Reality

James C. Rainwater, Daniel G. Friend, Howard J.M. Hanley, Allan H. Harvey,
Cynthia D. Holcomb, Arno Laesecke, Joseph W. Magee, Chris Muzny

Physical and Chemical Properties Division
Chemical Science and Technology Laboratory
National Institute of Standards and Technology
325 Broadway
Boulder, Colorado 80305-3328
USA

Forum 2000, which addressed the present needs and priorities for thermophysical properties measurements, was held June 29, 2000, at the Fourteenth Symposium for Thermophysical Properties in Boulder, Colorado. Seven distinguished experts presented overviews of issues related to a wide variety of subjects, and three discussion periods were held. This report documents in detail those overviews and discussion periods. Topics included whether simulation can replace experiment, properties needs for new miniaturization technologies, urgent problems such as nuclear waste cleanup, data needs for electrolyte systems and new generations of electric power plants, and data needs for unconventional materials such as molten metals and soft solids. Also included in this report are invited essays on issues raised at the Forum by panelists, audience participants, and other experts in thermodynamics.

Key words: Electrolytes; Forum 2000; miniaturization; nuclear waste cleanup; power plants; simulation; soft solids; thermophysical properties.

1. Introduction

The roles and priorities of experiment and simulation, as well as theory and correlation, for thermophysical properties research in upcoming years were the topics of Forum 2000, held as part of the Fourteenth Symposium on Thermophysical Properties at Boulder, Colorado. The Forum took place on June 29, the Thursday afternoon of the week-long symposium. All participants in the symposium were invited and encouraged to attend the Forum, which was organized by a committee of scientists, the authors of the present report, from the Physical and Chemical Properties Division of NIST, Boulder. The Forum brought together a panel and audience of distinguished authorities from academic, government, and private research institutions and funding agencies.

The advance abstract for the Forum, written by the committee, read as follows: "*Advances in miniaturization, decentralization, demand-controlled production, flexible feedstocks, and information technology will catalyze dramatic changes in the fluid-based industries in the 21st century. New technologies are emerging in areas such as waste minimization, advanced fuels, modular power plants, and high-value chemicals. Accelerated design, evaluation, and optimization of these processes require virtual tools based on robust information. Essential to these tools are physical property models, which must be validated with accurate data.*

"*All stakeholders in technology development reap the benefits from accurate measurements and improved property models. However, economic realities prevent single entities from committing substantial resources to such research. This Forum will identify strategic needs for collaborative efforts between experimentalists and developers of database and process modeling tools with direct input from the end users to respond to their fluid property needs. The intent is to bring together competence from industry, academia, and government research with representatives of the funding organizations to assist in the realization of these collaborative efforts. These efforts will result in a stronger connection between virtual design tools and physical reality.*"

The present report is a detailed documentation of the Forum. A condensed summary of the Forum has been published in the Journal of Chemical and Engineering Data,[1] and an article in Chemical Engineering Progress elaborating on some themes from the Forum is also in press.[2] Further information is available at the Forum website `http://Forum2000.Boulder.NIST.Gov`.

The Forum was moderated by Howard Hanley of the Physical and Chemical Properties Division of NIST, Boulder, Colorado. Each of the seven panelists presented ten-minute talks. In the first half of the Forum, there were discussion periods after the second and fourth talks, and then an intermission. The second half consisted of three additional talks by the panelists, a sum-

---

[1] **Rainwater, J. C., D. G. Friend, H. J. M. Hanley, A. H. Harvey, C. D. Holcomb, A. Laesecke, J. Magee, and C. Muzny (2001).** *Forum 2000: Fluid Properties for New Technologies, Connecting Virtual Design with Physical Reality.* J. Chem. Eng. Data, **46**(5), 1002-1006.

[2] **Harvey, A. H. and A. Laesecke (2002).** *Fluid Properties and New Technologies: Connecting Design with Reality.* Chem. Eng. Prog., in press.

mary from the moderator built around bullets for key topics of the Forum collected from members of the committee who reported on the proceedings, and a final and longer discussion period that concluded the Forum. The presentations were videotaped and, from the tapes, transcripts were made. The following sections are based on those transcripts, with direct quotation and paraphrasing of the speakers and audience participants. The panelists had an opportunity to review their sections and to edit them for clarity. The figures in the following sections are a representative selection of the viewgraphs presented by the speakers. The complete presentation files of the panelists are available at the Forum web site.

We also invited panelists, audience participants, and other authorities in thermodynamics to submit essays on issues raised at the Forum. Thirteen essays were received, which are presented in Section 14.[3]

---

[3] Papers submitted to this publication by non-NIST authors may have been edited by NIST, but such editing was done only to improve clarity of expression. Any comments or recommendations offered by non-NIST authors are those of the authors alone, and should not be in any way construed as expressing any official policies of, or recommendations by NIST.

2. Molecular Modeling

Peter Cummings

The first panelist to speak was Professor Peter T. Cummings of the Department of Chemical Engineering, University of Tennessee, Knoxville, and Oak Ridge National Laboratory. Cummings first mentioned that Howard Hanley, when inviting him, had asked him to be provocative and that he would try to be. He said that he had largely composed his talk in the three days since Professor William A. Wakeham's plenary lecture at the symposium, titled "A Possible Future for the Thermophysics of Fluids." Cummings stated that his point of view was from "somebody who does a lot of molecular simulation." He mentioned the movie 'The Matrix' as an example of the influence of simulation. "In 'The Matrix,' the whole idea is that we live in one gigantic computer simulation." He joked that "if 'The Matrix' were true, there would be no true experimental data: All data would be the result of simulation!"

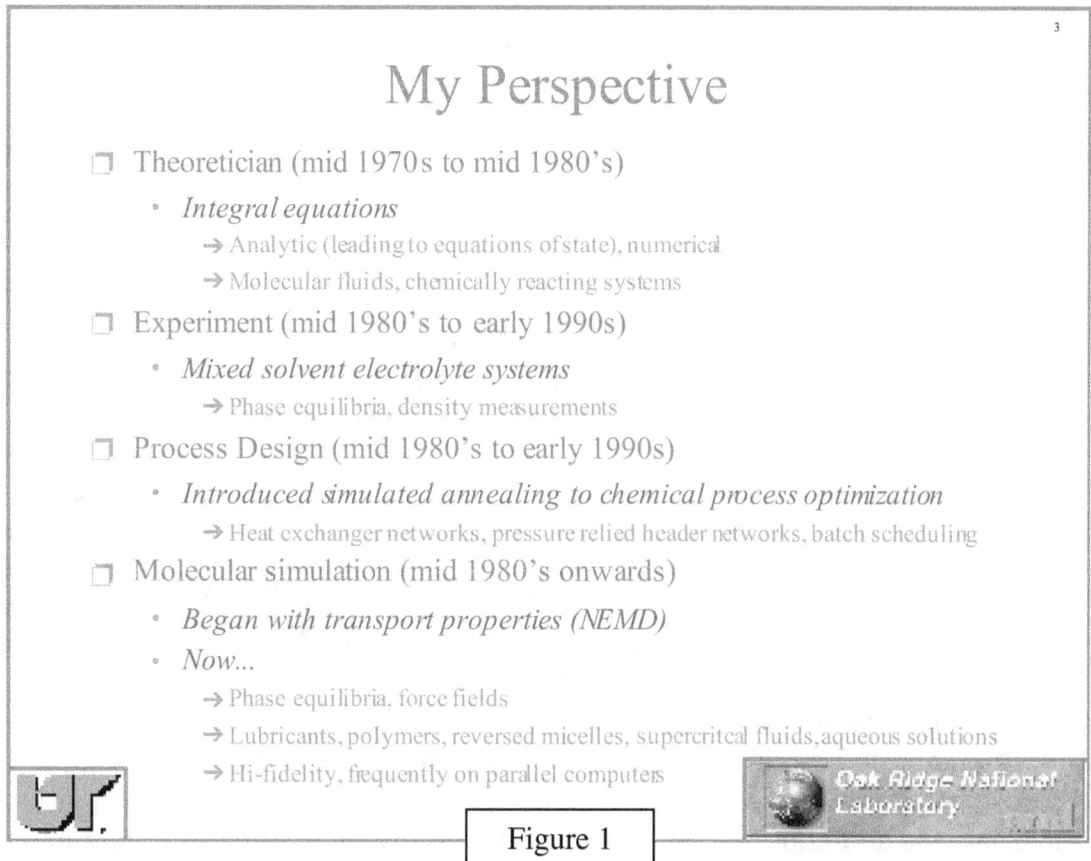

Figure 1

Cummings described his extensive background in many aspects of the profession (Fig. 1). "I'm coming from the perspective of somebody who has been a theoretician. In fact, when Bill was talking Monday about simulation squeezing out theory, I was thinking about my career in that respect. I used to do theory up until the mid-1980's, mostly integral equations, particularly analytical solutions of those. I did some experimental work over almost a decade on mixed salt and electrolyte systems, so I have some feeling for (admittedly unsophisticated) experiments. I worked on process design over a period of about seven years, where we introduced simulated annealing to the chemical process design community and designed things like heat exchanger networks, pressure release header networks, doing back scheduling of plants and so on. In the last 15 years or so, molecular simulation has been a major focus in my group. It began with transport properties, but we've expanded our focus to include areas such as phase equilibria. Applications to lubricants, polymers, reverse micelles, and supercritical fluids have forced us into the development of force fields. I call these simulations high-fidelity simulations, where we try to match experiments, not try to match theory, because we really are trying to predict the properties of systems directly, and frequently we do this on parallel computers."

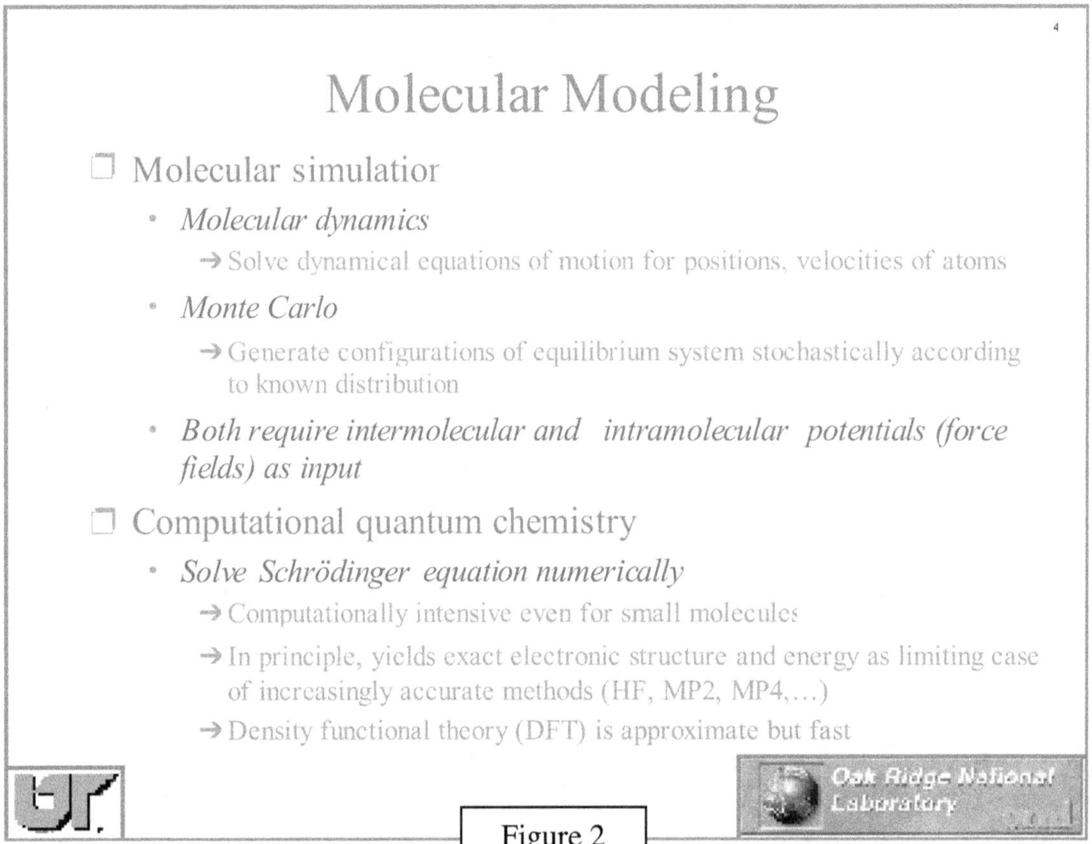

Molecular Modeling

- Molecular simulation
  - *Molecular dynamics*
    → Solve dynamical equations of motion for positions, velocities of atoms
  - *Monte Carlo*
    → Generate configurations of equilibrium system stochastically according to known distribution
  - *Both require intermolecular and intramolecular potentials (force fields) as input*
- Computational quantum chemistry
  - *Solve Schrödinger equation numerically*
    → Computationally intensive even for small molecules
    → In principle, yields exact electronic structure and energy as limiting case of increasingly accurate methods (HF, MP2, MP4,...)
    → Density functional theory (DFT) is approximate but fast

Figure 2

Cummings emphasized that there are two different aspects of molecular modeling (Fig. 2). "I use that term to describe a couple of things—one is molecular simulation, meaning molecular dynamics, where you solve equations of motion of the molecules, and get properties of the system, and Monte Carlo, where you do it stochastically. Both of those require intermolecular and intramolecular potentials, or their force fields, as input. The other part of molecular modeling is computational quantum chemistry, where you solve the Schrödinger equation numerically for, typically, the electronic degrees of freedom of a small number of atoms. It's computationally intensive even for small molecules. There have been a lot of papers at this meeting that included computational quantum chemistry, and also included things like density-functional theory, which is one of the recent developments that is making it more computationally efficient."

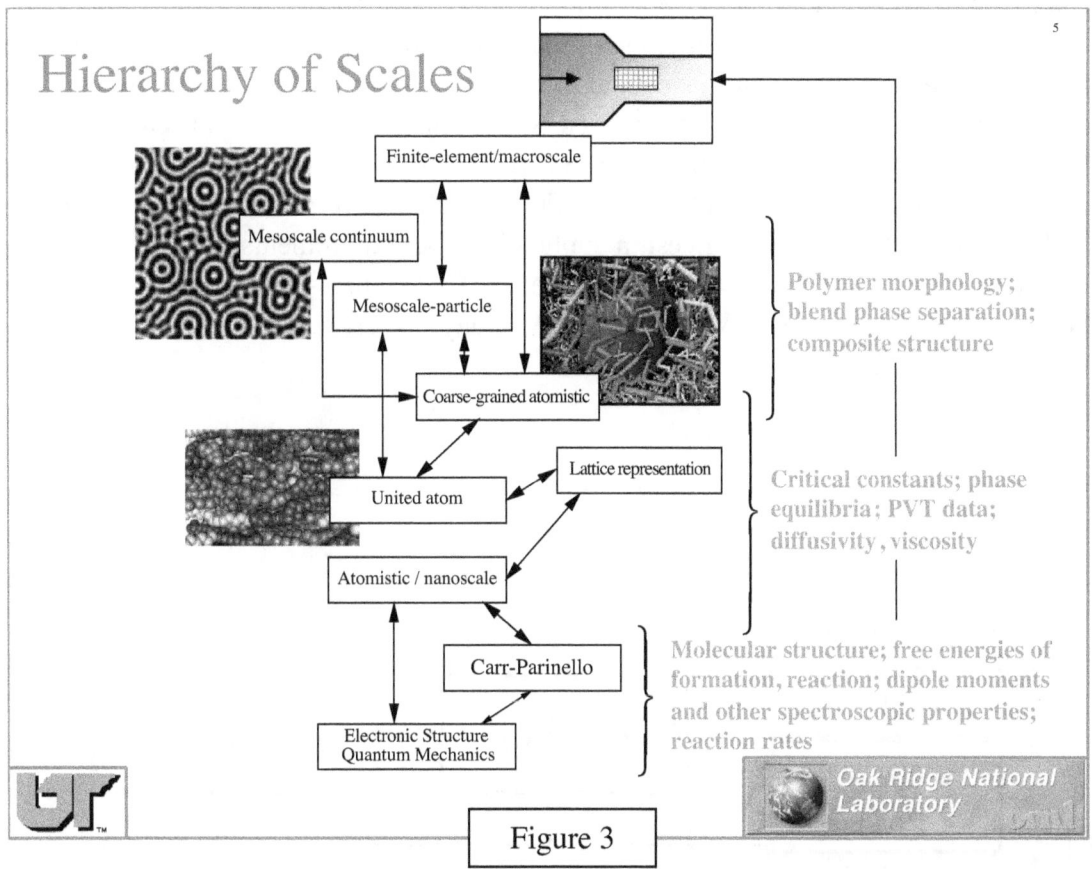

Figure 3

He next specified the various length scales within molecular modeling (Fig. 3). "You go from the lowest level, the electronic structural quantum chemistry scale, through scale-bridging methods like Car-Parrinello that get you up to the atomistic scale. At the lowest level, quantum chemistry, you can get things like molecular structure, free energies of formation and reaction, enthalpies of formation, dipole moments, and other spectroscopic properties which then can be used at the highest level, which is the continuum-mechanics and process-design level. From the molecular simulation techniques, you can get things like critical constants, phase equilibria, PVT-type data, transport properties such as diffusivity and viscosity. A lot of people nowadays work in the area of polymers, where you're looking at things like polymer morphology, polymer blends, separation, rheology of polymers." He mentioned that, with polymers, the relaxation times were so slow that it was difficult to obtain properties that go directly into process simulation, because what happens at the process level affects what happens at the molecular level.

Cummings pointed out that there are two foundational areas of research needed for molecular simulation, methods and force fields, and that some people work on one or the other and some on both (Fig. 4). "The force fields typically come from *ab initio* calculations as well as experimental data; things like dipole moments, polarizability, and so on. But, it also happens that as methods improve, you can then use additional experimental data to parameterize the force fields that are used in the calculations." He noted that, historically, force field development has been driven by the design of pharmaceuticals, rather than chemical processes involving fluids. "It's clear that the degree to which drug design is integrated into the pharmaceutical companies and the amount of money that's being spent on research in that area dwarfs dramatically the amount of money that is being spent on force field development and simulation techniques for chemical processes. And in the drug design application, the requirements in terms of state conditions are very modest. The temperatures are 25 to 35 °C. The pressures are typically less than 5 bar. And, as we heard from Bill [Wakeham] on Monday, experimental data in that range won't even make it into his journal."

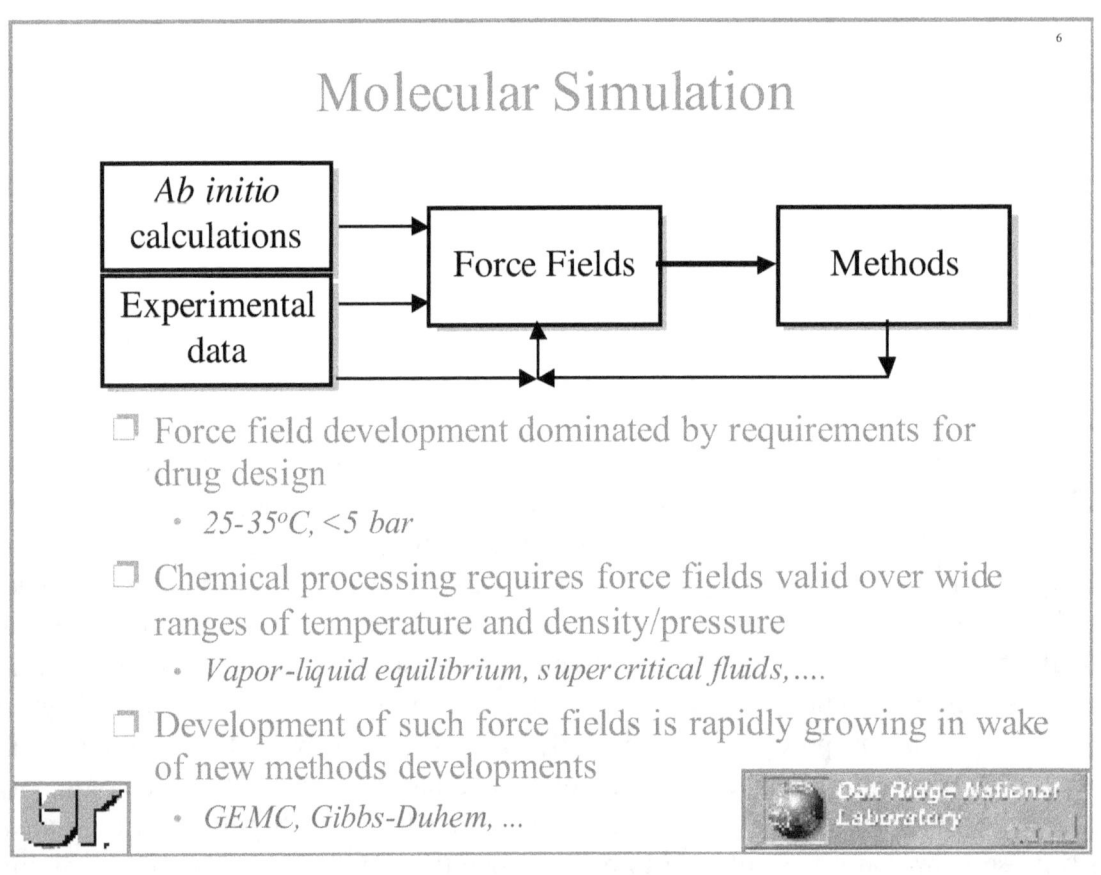

Figure 4

> # Molecular Simulation *vs* Theory
>
> ☐ Advances in computational hardware and algorithms
> - *Moore's law*
>   → Computing speeds double every 18 months = order of magnitude every 5 years
>   → Add 2-3 orders of magnitude from parallelization (cheap today)
>   → Costs driven by consumer market
>
> ☐ Costs for experiment?
> - *Labor-intensive, high capital costs*
>
> ☐ Costs for theory?
> - *Labor-intensive^2*
>
>   *Do graduate students and/or lab personnel/equipment improve by an order of magnitude every five years?*
>
>

Figure 5

By contrast, Cummings noted, chemical processing requires force fields valid over wide ranges of pressure, temperature, and density (Fig. 4). "You just have to think about vapor-liquid equilibrium, where typically the densities can range by an order of magnitude or more between the vapor and liquid phase, and the temperature range that you're looking at can vary over hundreds of degrees Celsius, which is certainly very different from what's required in the biological applications. The development of force fields that will work for chemical processes is one of the rapidly growing areas in molecular simulation, and it's grown in the wake of development of techniques that enable you to calibrate potentials so that they can work for phase equilibria. Things like the Gibbs ensemble Monte Carlo method of Panagiotopoulos, the Gibbs-Duhem method of Dave Kofke, and so on."

As for the dramatic growth in the field of molecular simulation and its increasing displacement of traditional theory, Cummings mentioned Moore's law as a contributing factor (Fig. 5). "Computing speeds double every 18 months. That's Moore's law. The way I prefer to think about it, is an order of magnitude every five years. Since I finished my Ph.D., computing has increased by four orders of magnitude. You can add on top of that parallelization, which can add you two or three orders of magnitude when you put together 100 or 1000 processors. And the big thing about that is that you get tremendous leverage out of the consumer market for computers. This laptop that I have here has a peak processing power of 800 megaflops which, not too long ago, in the early to mid 90s, was what you got with a four-headed Cray Y-MP." He also gave examples of the needs for word processing and computer games driving companies to make more powerful computers.

Molecular Modeling

☐ International comparative study on applying molecular modeling
- *Goal to evaluate ways in which molecular modeling being applied, primarily in industry, throughout the world*
- *US funding agencies*
  - → NSF, DOE, NIST, DARPA, AFOSR, NIH,...
- *Over 75 sites visited worldwide*
  - → Primarily companies, mostly in Europe, Japan and US
- *Report to be published in late 2000*
- *Web site*
  - → `http://www.itri.loyola.edu/molmodel`

 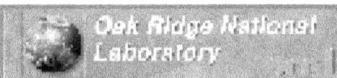

⇑ Figure 6   Figure 7 ⇓

Molecular Modeling

☐ Three main roles:
- *Predicting fundamental properties used in engineering correlations*
  - → E.g., critical constants, molecular structure, dipole moment
- *Predicting required properties directly*
  - → E.g., phase equilibrium of mixture
- *Providing conceptual molecular-level understanding of properties*
  - → E.g., developing correlations, evaluate theory, guide/supplement/replace experiment

☐ Additional roles
- *Intellectual property protection (defense and offense)*
- *Development of QSAR/QSPR for product design applications*

 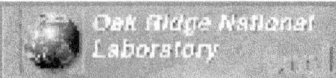

He gave reasons why such developments put traditional research methods at a disadvantage (Fig. 5). "The price for experiment is that it's labor-intensive; it has high capital costs. The cost for theory is even more labor intensive And, so part of the reason I think why people are doing simulation more is that graduate students and lab personnel or equipment do not appear to improve by an order of magnitude every five years. I may be wrong. I'm not doing that much experiment or theory these days." He mentioned a survey, conducted at the behest of various Federal agencies (Fig. 6), with a website, of the way molecular simulation is being applied throughout the world. Molecular modeling is not just being used for the expected tasks (Fig. 7) but also in courtrooms for intellectual-property defense in connection with patent law.

He mentioned that Wakeham had reviewed how, over the years, research in thermophysical properties had become increasingly oriented toward chemical engineering and process engineering requirements. Cummings argued that the trend reflected where people in the United States are conducting fluids research (Fig. 8). "A lot of the people who do fluids are in chemical engineering; a lot of people who've attended this meeting first attended this meeting not as chemical engineers, and came into this later on."

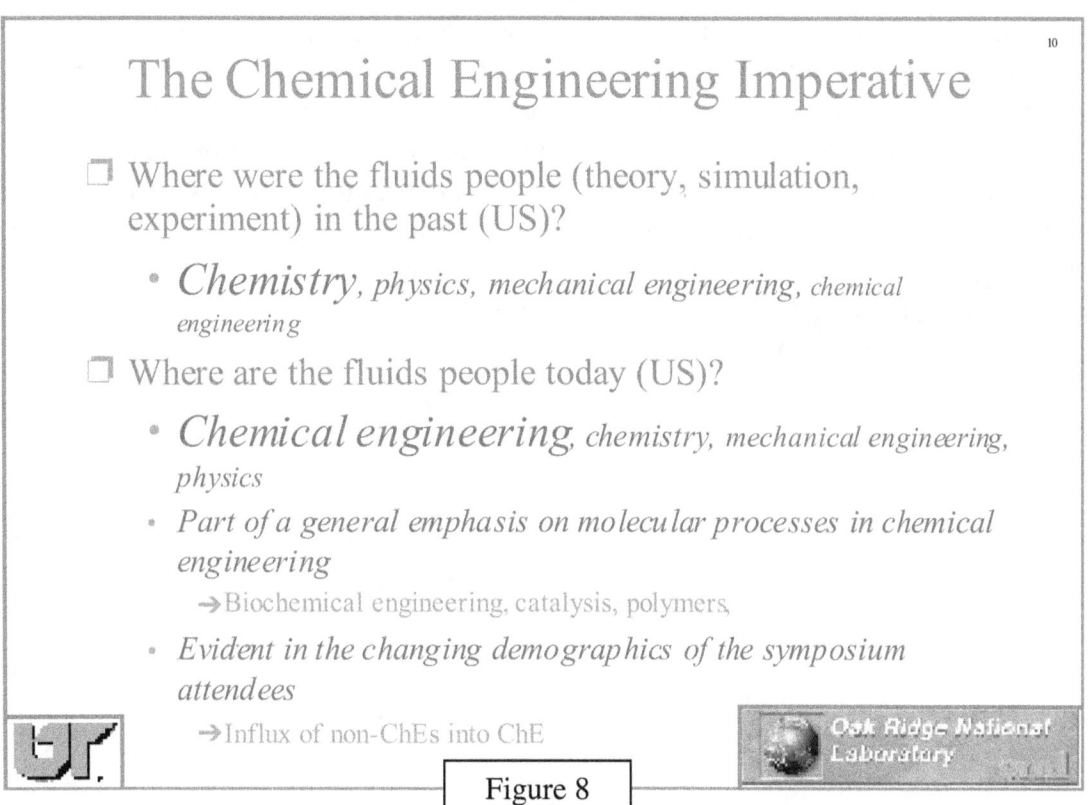

Figure 8

Cummings described the accuracy requirements for the early stages of process design and why computational techniques are frequently useful here (Figs. 9-10). As a final example, he mentioned instances where some companies now rely on computational results (Fig. 11). "One of the companies we visited now routinely gets all its enthalpies of formation and enthalpies of reaction from computational chemistry. Five years ago, an experiment cost them $50,000, computation $20,000. Today, an experiment costs over $100,000, computation costs $5,000, and this cost drops by a factor of two every 18 months. And, the accuracy of the computer properties, they believe, is as reliable or more reliable than experiment." Some companies that are using these techniques are listed in Figs. 12 and 13.

Cummings concluded his presentation with some perspectives and predictions regarding the role of molecular simulation (Fig. 14). "I think by 2020, it's going to be routine that many of the thermophysical and thermochemical properties of low-molecular-weight systems will be predictable computationally, and at better accuracy and higher precision than experiment. Within the simulation community, there's an increasing emphasis on trying to predict things that haven't yet been measured. The way we've typically validated things is to predict the values of quantities that have already been measured experimentally, and that always leads to the criticism: 'for this one you published where it did agree with experiment, how many did you do where it didn't.' And, so, I think there's an encouragement of a greater mode of prediction of properties that aren't already measured."

"I believe theory and computation will continue to be essential for systems that are challenged by simulation, systems with large relaxation times, systems whose properties are determined at scales larger than the molecular, mesoscales; these include materials where quantum physics or quantum chemical processes are important. Also, I have to point out that progress in simulation is only going to be possible with new theories for bridging time scales, and if anybody is interested in that later on, I can talk about that. I don't think theory is gone by any stretch of the imagination, or experiment is gone by any stretch of the imagination, but molecular modeling techniques are becoming very powerful and very important players in the whole area of thermophysical properties prediction."

Process Design Overview

- Three stages (Zeck and Wolf, 1993; Douglas, 1988)
  - *Stage I: Process screening*
    - →Elimination of most alternatives
      - Only ~1% of proposed chemical processes result in commercial production
      - Use of shortcut methods for design calculations (material and energy balances)
      - Modest data accuracy requirements (total cost accurate to ±25%)
      - 90% or more of designs eliminated using these methods
  - *Stage II: Process development*
    - →Detailed economic assessment of several process alternatives
  - *Stage III: Process design*
    - →Detailed design and optimization of chosen process

⇑ Figure 9   Figure 10 ⇓

Properties Required for Process Screening

- Physical properties frequently estimated by engineering correlations or measured by simple, low-cost experiment (e.g., infinite dilution activity coefficients by gas chromatography)
- Accuracy of ±25% acceptable in cost estimates
- Demands for data accuracy vary (Larsen, 1986)
  - *20% error in density --> 16% error in equipment size/cost*
  - *20% error in diffusivity ---> 4% error in equipment size/cost*
    - → Errors in density usually small for liquids, errors in diffusivity frequently large (factor of two or more)
  - *10% error in activity coefficient results in negligible error in equipment size/cost for easily separated mixtures, but for close-boiling mixtures (relative volatility <1.1) 10% error can result in equipment sizes off by factor of 2 or more*
- Thermophysical properties data must be accompanied by accuracy assessment

Potential Impact of Molecular Modeling on Process Design

- Provision of physical properties data at Stage I and possibly Stage II
- Guidance for experimental studies at Stages II and III
  - *What are the troublesome mixtures and/or state conditions?*
- Dual role
  - *Provide raw physical properties "data" required for correlations (indirect)*
  - *Provide directly properties of pures and mixtures*
- Example from thermochemistry - enthalpy of reaction
  - *"Now in maintenance mode"*
    - → Five years ago: Experiment cost $50,000, computation $20,000
    - → Today: Experiments cost over $100,000, computation $5,000

 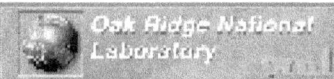

⇑ Figure 11   Figure 12 ⇓

Computational Chemistry

- Industrial Case Histories

Air Liquide	Design zeolites for O_2/N_2 separation
Air Prod & Chem	Adhesives, adsorption
Albemarle	Flame retardancy
Amoco	Catalysis-homo/heterogeneous thermochemistry
BP	CH_4 adsorption evaluation
Chevron	Gas hydrates, lubricants
Dow	Thermochemistry, reaction networks, ΔH_{rxn} (safety)
DuPont	Thermo, kinetics, catalysis
Exxon R&E	NO_x kinetics, elementary and networked reactions, safety

- *Compiled by Phil Westmoreland, UMass*

 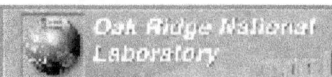

Computational Chemistry

☐ Industrial Case Histories

Hercules	Polysaccharide rheology
Lubrizol	Hydrocarbon-chlorine reactions, anti-corrosion additives
General Motors	Monolithic catalysts
Procter & Gamble	Designed detergent enzyme
Royal Dutch Shell	Diffusion in porous media
Schlumberger	Setting of cements
Shell USA	Solvent separations, catalysis
Union Carbide	Catalysis, materials
Xerox	Modify and develop materials

 • *Compiled by Phil Westmoreland, UMass*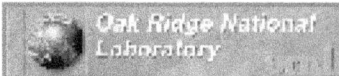

⇑ Figure 13   Figure 14 ⇓

The Future

☐ By 2020, many routine thermophysical and thermochemical properties of low molecular weight systems will be predictable computationally at better accuracy and higher precision than experiment
☐ Increasing emphasis on prediction by simulation
☐ Theory and experiment will continue to be essential for systems challenged by simulation
  • *Long relaxation times (polymers, glasses)*
  • *Materials whose properties are determined at scales larger than molecular (mesocale)*
  • *Materials in which quantum physical and/or chemical processes are important*
☐ Progress in simulation is possible only with new theories for bridging time scales

 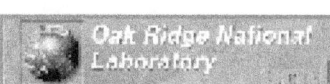

3. Microchemical and Thermal Systems: Process Intensification Through Miniaturization

Ward TeGrotenhuis

The next panelist was Dr. Ward TeGrotenhuis, Chief Engineer, Environmental Technology Division of the Pacific Northwest National Laboratory. He spoke about microtechnology, a topic of particular interest to Forum 2000 in the search for applications of thermophysical properties to new and emerging technologies. The title of his presentation was "Microchemical and Thermal Systems: Process Intensification Through Miniaturization."

Figure 15

New Tools: MicroChemical and Thermal Systems

- Microchannel Heat Exchangers, Reactors and Separation Units
- Miniaturization:
  - Process Intensification
  - High capacities
  - Lightweight systems
  - Mass Production

DARPA:	MEMS, Mesoscopic Machines
NASA:	Micro/Nano Systems
Department of Energy:	Microcats

TeGrotenhuis presented viewgraphs of several miniaturized devices, starting with a microchemical heat pump minus the compressor (Fig. 15). He summarized the various general objectives in miniaturization technology. "We have built numerous devices, reactors, heat exchangers, and separation units, and the general idea is to achieve miniaturization through process intensification without sacrificing high capacity, and also achieving a lightweight system. That's the general idea. A comment that we typically get is that these devices look way too expensive to ever be economical, and the argument we pull up is from an analogy of the microelectronics industry, where we are looking for economies of mass production as opposed to economies of scale. There are a number of U.S. agencies now that have very active programs in this area, including DARPA, NASA, and the Department of Energy."

He displayed the length scales involved in microtechnology on a logarithmic axis (Fig. 16). "We typically are working with microchannel widths that are down around the micron scale up to perhaps a millimeter. This leads to pumps and valves that can be, maybe, up to about a centimeter in scale, reactors and heat exchangers that can be up to 10 centimeters. And then an overall thermal or chemical system that is perhaps a meter to slightly more. This is in comparison to conventional chemical process equipment where pumps and valves and systems are typically two orders of magnitude or more in size."

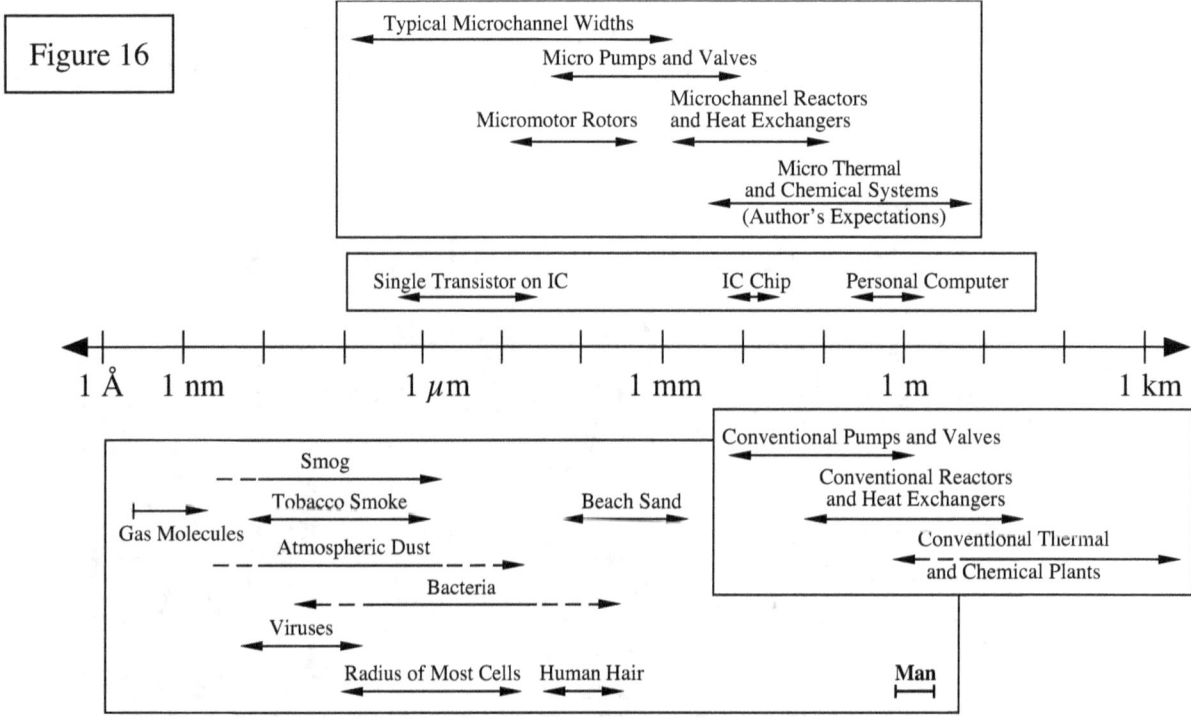

TeGrotenhuis next gave an example of what he called one of their more notable success stories (Fig. 17). "This is a device that won an R&D 100 Award last year. It's a gasoline vaporizer for an automotive fuel processor. It's designed to handle a process for vaporized gasoline at a 50-kilowatt capacity. It has four reactors and four heat exchangers built into it. Its total volume is about a third of a liter, and it processes about 1400 standard liters (vapor phase) per minute of gasoline. This is over an order of magnitude smaller than any other technology that's available to do this function."

He described the very rapid heat and mass transfers that can be achieved (Fig. 18). "For microchannel heat exchangers, we get some statistics here. Heat fluxes, we can exceed 100 watts per square centimeter of surface area. Another comment we typically get is that looking at these very small channels, the pressure drops will be way too high. But, through the massive parallelization and very short channels, we're able to minimize the pressure drop down to the order of a psia [6.9 kPa] or so. We're able to achieve heat transfer coefficients on the order of 10,000 to 15,000 watts per square meter per kelvin for liquid, and for the evaporated phase, 30,000 to

35,000 is achievable. This picture down here shows these channels are roughly 250 microns wide and a couple of millimeters deep."

Some further specific examples of devices were presented, beginning with an under-the-hood automotive steam reforming system (Fig. 19). "A number of European companies as well as the U.S. Department of Energy are very active in trying to develop fuel cell systems for automobiles. And one approach to providing the hydrogen is to do fuel reprocessing under the hood, where you would take a liquid fuel such as gasoline and reform it to produce the hydrogen-rich gas stream. Now, most people who approach this problem are using partial oxidation or autothermal because the kinetics are fairly proven to be fast. And steam reforming is generally avoided because it's believed to be a slow reaction and requires a very large reactor. But we've demonstrated that we can do steam reforming on the millisecond time scale. So we've been pursuing steam reforming, and because it's an endothermic process, there's a number of heat exchangers and vaporizers that are networked together to achieve the thermal efficiency for the system."

Gasoline Vaporizer

- 50 kWe capacity: Four cells each of microchannel reactors and heat exchangers
- Volume: 0.3 liters
- Processes/combusts 1400 SLPM

1999 R&D 100 Award Winner

Figure 17

He then showed the microchemical process network for testing the fuel processing (Fig. 20). "It includes a number of heat exchangers. There are the reactors down at the bottom you can't quite see, but each of these other little boxes are heat exchangers as well as the ones on the

Advantage: Rapid Heat and Mass Transport
Microchannel Heat Exchangers

- High convective heat transfer coefficients
- Heat fluxes: >100 W/cm^2
- Low pressure drops: 1-2 psia

- Liquid phase: 10,000 - 15,000 W/m^2·K
- Evaporating phase: 30,000 - 35,000 W/m^2·K

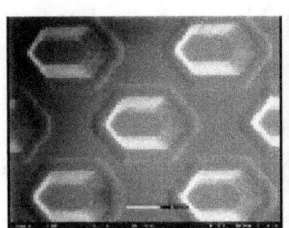 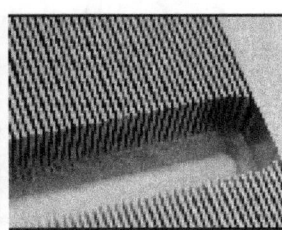 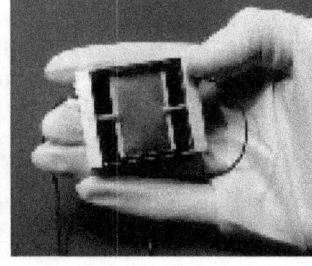

Figure 18

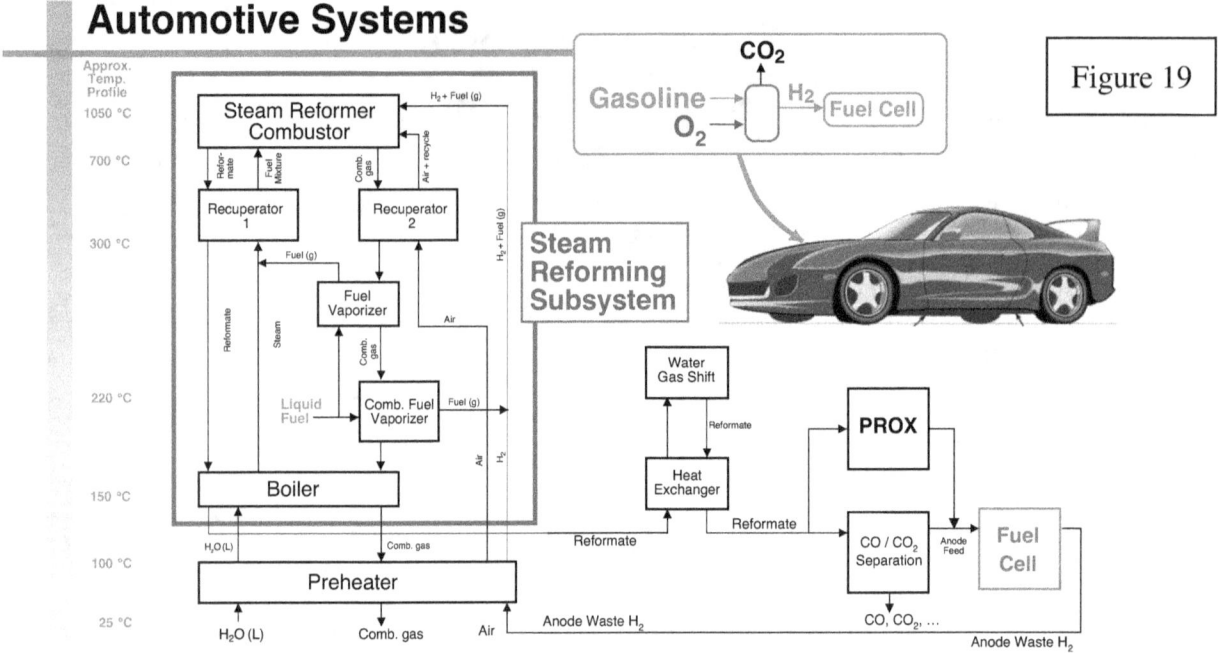

Figure 19

outside. There are four parallel process streams built into this single unit. There's a lot of tubing on there, but it's all there for thermocouples and instrumentation. Eventually, all of these are designed so they can collapse down into a single device. This represents roughly 20 kilowatts equivalent, or roughly about half scale."

The next example he presented was a device built for NASA, for *in situ* propellant production on Mars (Fig. 21). "NASA intends to send chemical processing plants to Mars to produce propellants from the CO_2 that's present in the atmosphere. We have a program with them to develop microtechnology components for an *in situ* propellant production plant. These are both artists' renditions for an ISTP plant for a sample return mission, and then this is an overall facility for human exploration. The only way that human exploration is really affordable on Mars is to develop or make the propellants for the return trip from the CO_2 in the (Martian) atmosphere."

Figure 20

He mentioned two additional projects. The first was for the Department of Defense, the development of a portable device for cooling a soldier in the field (Fig. 22). The second was the

NASA: In Situ Propellant Production on Mars

Figure 21

development of systems considered as a "platform" for nanotechnology. "We're developing engineered structures that have improved heat and mass transfer for catalysis as an example. We also are developing self-assembling surfaces for controlled wettability and surface properties. Also, we're getting into nanofabricated structures. This micrograph (Fig. 23) is platinum clusters on a substrate. And we're also working on biological enzymes as highly functional catalysts. And all of these, we believe, can very naturally fit into the microchannel architecture to convert to a usable process."

The fluid thermophysical properties issues, both traditional and nontraditional, with microtechnology were of particular interest to Forum 2000, and TeGrotenhuis addressed them in his concluding viewgraph (Fig. 24). He described the design development to date as "Ready, fire, aim," that is, trying various possibilities without detailed fundamental analysis beforehand, and within this approach, fluid properties have not been of great importance. "But, we see as these things get closer and closer to actual development, we're now beginning to actually design devices for outside entities. And it's becoming more and more important that we have very good property data."

"Some of the issues that have come up: First of all, the behavior of fluids at this scale. Certainly, from the devices I've shown you, the wall effects are very important. They dominate in these systems. We're constantly dealing with surfaces and interface-type issues, including fouling. When you talk to a process engineer, they can hardly imagine putting a dirty process stream, say in a chemical plant, through one of these small channels. So, fouling is certainly a very large concern."

DOD: Man-portable Power & Cooling

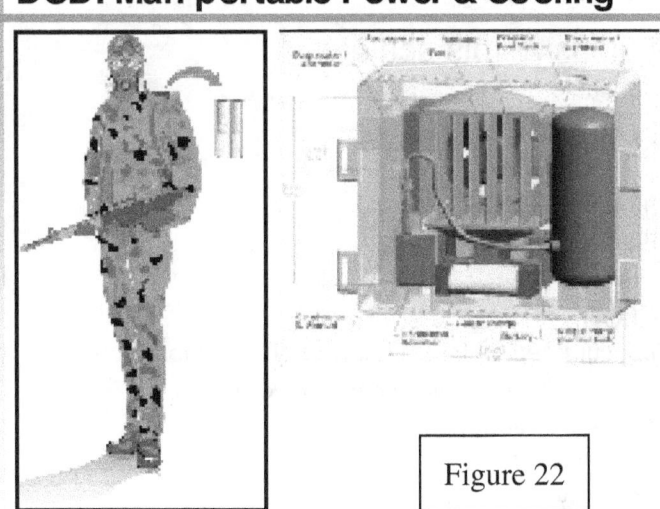

Figure 22

TeGrotenhuis concluded with some further issues, including the instrumentation needs for basic state measurements, as well as the need for traditional properties such as densities and thermal and electrical conductivities as design parameters. Controls were mentioned as important. "We know from nature that as you reduce the size scale of a process, typically the time scale reduces accordingly. What we're finding is that conventional control systems simply can't handle the dynamics that are happening at this scale. Fluid flow and distribution and the general system homogeneities are important. The ability to measure properties in these channels, and actually characterize what's happening, is something that I think will advance this technology tremendously."

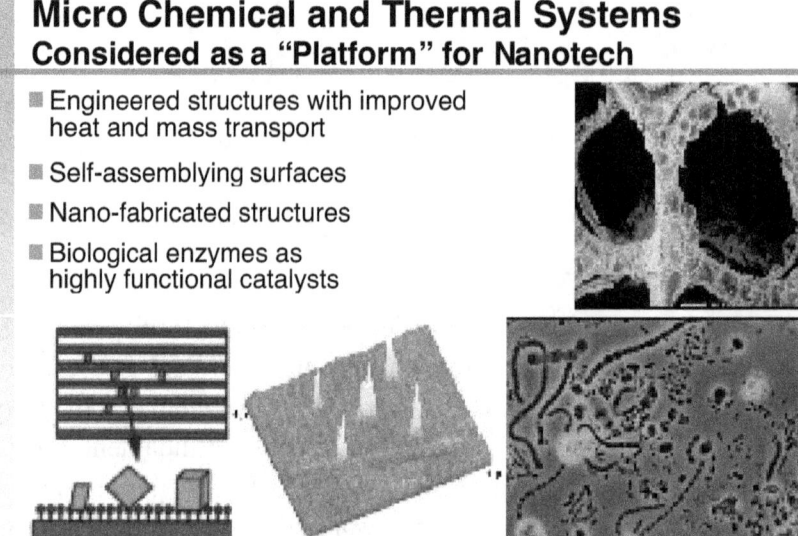

Figure 23

Micro Chemical and Thermal Systems
Considered as a "Platform" for Nanotech
- Engineered structures with improved heat and mass transport
- Self-assemblying surfaces
- Nano-fabricated structures
- Biological enzymes as highly functional catalysts

Figure 24

What are the property issues?
- Behavior of fluids at this scale
  - Wall effects dominate
- Surfaces and interfaces
  - Fouling
  - Multi-phase flow
  - Surface energy, tension, wettability
  - Dynamics -- start-up
- Instrumentation
  - Basic state measurements - P, T, phase, quality, composition
  - Density, thermal conductivity, electrical conductivity
  - Controls -- need to be integrated
  - Fluid flow and distribution
  - System homogeneity

4. First Discussion Period

After the presentation of TeGrotenhuis, moderator Hanley called for the first discussion period, to be focused on the first two talks. In this report, audience participants and their affiliations are listed in Table 2.

Siepmann offered an addition to Cummings' talk. "Speaking for the people who develop algorithms, I would strongly argue that every one of us would prefer a ten-year old computer with today's algorithm than the other way around. So that means that the speed of improvement coming from novel algorithms far exceeds the speed improvement just alone from the computers. So, I would argue that we most likely go up by something like four or five orders in magnitude per decade." Cummings replied: "I agree completely with what you said, Ilja [Siepmann]. And at the end I pointed out that a real important role of theory in future is in the algorithms for molecular simulation as opposed to theory on the physical systems themselves."

Thompson posed a question about miniaturized laboratory equipment. "Do you foresee a bank of five cubic centimeters vapor-liquid equilibrium cells with automatic filling, automatic sampling to a GC. Perhaps five minute measurements. Does that sort of thing appeal to those of you that make VLE measurements?" Wakeham replied in agreement. "We're already moving a long way in that direction. Of course it wouldn't have been possible a few years ago to think of that."

Hanley asked a question of TeGrotenhuis."You gave a list of property information that was needed, but is this property information associated with the instrument or is it sort of general property information?" TeGrotenhuis replied that until now his group had looked only at traditional property measurements. "It's not necessarily associated with the scale. So we see pretty good agreement with the calculated predictions for the performance of the devices with conventional bulk property measurement." Hanley countered: "That must eventually break down, though." TeGrotenhuis replied: "Yes. We're really in a technology development mode where we don't see the effects at this point. I anticipate eventually we will see the effects of non-idealities associated with the scale."

Laesecke brought up a point from one of Cummings' slides that was not shown on critical points of n-alkanes as determined by simulation. "This example is often quoted for the decisive impact that simulation can have to determine the correctness between two experimental methods. But, if you read the erratum two years later, you found that the original simulation had a great error in itself, a systematic error, and the critical temperatures from the simulations had to be corrected by a great deal, by tens of kelvins. This correction is never quoted." He said experimentalists would never be allowed to present such inaccurate data.

Cummings said that Siepmann might best reply about that specific work, but commented: "I believe validation is best when simulationists predict properties that then are subsequently measured. Part of the difficulty we face when we're developing force fields is that we have to fit to a certain amount of experimental data and in the situation mentioned (and Ilja may want to comment on this with respect to that particular calculation) but I think that there was some inconsistency with the long-range correction."

Cummings then deferred to Siepmann, who replied: "You are saying that the simulations were wrong. I could argue equally well, the only thing that was different between the erratum and the final publication is that we missed out on the tail correction. So if you use exactly that same force field without the tail correction, you still get the old answer back. So nothing has changed on the simulation. There was no bug in the code or anything." In terms of the publicity of that paper, he said: "The only emphasis on that publication and why it maybe got hyped so much is, there was a disagreement between experimental studies whether the critical density as a function of chain length would go through a maximum and then go down. In particular, Amyn Teja predicted that it would go down after going through a maximum, and his results were not believed. And there was some additional study at NIST,[4] which predicted that the critical density would go up as a function of chain length, and Flory-Huggins theory, for example, predicts that it goes down. And I think what the simulation has clearly shown is that it goes down and I think there is no doubt about it and that hasn't changed. And I think present-level force-fields seven years later can now predict these experimental properties with higher accuracies than experiment can."

Kincaid returned to Cummings' comment about the movie 'The Matrix.' "Computer simulation is just at the stage that weather prediction was at a few years ago. You can predict the weather very well, locally, now using computational models which you might call *ab initio* with boundary conditions. But the real weather prediction accuracy comes from integrating those results with a huge data base of what the weather was like fifty years ago under these same conditions." He argued that prediction of thermophysical properties would evolve similarly. "In this sense experiment is a sophisticated kind of simulation of the real world and computer simulation is just a different kind of simulation and the two things have to be integrated tightly. And we have the computational resources, the database resources and the analysis skills now, to really start that integration and I think that's, maybe I'm wrong, but I tended to interpret Peter's remarks as that we're headed in that direction, not where some *ab initio* quantum chemistry calculation is going to tell us exactly what happens, because I think that's ridiculous. That's like trying to solve the Navier-Stokes equation for the world to predict the weather tomorrow in Denver. It's just not going to happen."

Cummings agreed, and offered a further example of a prediction from a DIPPR database (of the viscosity of perfluorobutane).[5] "It turns out that this correlation was quite wrong and that

---

[4] Ed. note: The study was actually carried out at NIPER and the results were quoted in **Tsonopoulos, C. and Z. Tan (1993)**. *The Critical Constants of Normal Alkanes From Methane to Polyethylene II. Application of the Flory Theory.* Fluid Phase Equilib., **83**, 127-138.

[5] **McCabe, C., D. Bedrov, G. D. Smith, and P. T. Cummings (2001)**. *Discriminating between Correlations of Experimental Viscosity Data for Perfluorobutane Using Molecular Simulation.* Ind. Eng. Chem. Res., **40**(1), 473-475.

the error was discovered by people doing simulations of those systems. If you look at what people do today in terms of molecular simulation in predicting thermophysical properties, one way you can look at it is that it's an extremely sophisticated form of group contribution technique. We tried to develop the intermolecular force fields with parameterizations that are based on experimental data just like people have done with things like activity coefficient models with much simpler, in a sense, science in them than simulation has in it and what you find then is that these are much more robust over wider ranges of operating conditions than these much simpler models." Cummings predicted the eventual replacement of engineering correlations calibrated with experimental data by simulation techniques which are also calibrated with data but, he suggested, will be more reliable.

Mountain posed a question to TeGrotenhuis. "At what sort of length scales do you think that you are going to find that the extant data sets are not adequate." TeGrotenhuis replied that he might not have a definitive answer. "Right now we're typically working in 100 micron to 200 micron channels, and there when we compare heat transfer and mass transfer performance in the device to either correlated or based on fluid dynamics CFD type codes, we get very good agreement, so at that scale we don't see any need for any refinement, but I think certainly as you get more towards the nanoscale then things will begin to break down." Hanley asked TeGrotenhuis where he obtained his properties data, and TeGrotenhuis replied that they were obtained form basic sources such as the CRC Handbook and Chemcad.

Pellegrino posed a question to the simulators. "It would seem that the value of simulation would come mostly for those systems where there is danger or hazardous materials. Is there enough experimental data for those systems that you could validate your property simulations?" Cummings replied: "In the safety area is one of the places where companies are at least willing to talk about what they're doing successfully with computational techniques such as molecular simulation or quantum chemistry. Certainly, when you talk to people from various companies, it becomes clear that they're relying a lot more on quantum chemistry calculations for quantities such as enthalpies of formation. In the safety area I can see where those sorts of thermophysical data are very important and you want to get those right. There's a lot of examples, both informal and explicit, where people have found errors in experimental data through quantum chemistry calculations of those kinds properties. I don't have a good feel in my mind for areas in which molecular simulation plays such a role in the safety calculations." Cummings mentioned that there were tentative plans to organize a session at an upcoming AIChE meeting on applications of quantum chemistry in the area of chemical safety.

Rainwater proposed a hypothetical scenario to Cummings. "About 15 years ago or so the Montreal Protocol and the concern about the ozone layer and alternative refrigerants, you had a situation where there was a whole class of fairly simple fluids that had not been measured very much and desperately needed to be measured. Then there was a 10 year or so huge effort worldwide to measure them, more or less completed by now. But, let's say hypothetically, your scenario for computer calculations existed at that time. Would you have said, 'Don't worry about doing all this experimental data, we can just figure it all out on a computer' or would you have really said, 'We should have done just as much anyway. It was essential to our efforts to do that.' We don't have any idea what sort of future emergencies like that might occur. So, at what point are you guys going to say 'We still need data' or 'We don't need it anymore, because a computer can figure it out'?"

Cummings replied with a clarification. "The real immediate role, I think, for these computational techniques is providing properties at a point early in the design process, where you are really trying to eliminate alternatives of designs, rather than trying to do a detailed process design. I don't think anybody would build a plant or adopt a new product without at some point performing experiments on it. Why spend a quarter of a billion dollars building a chemical plant if you haven't done any experiments at all on the systems that you're trying to design equipment for? Early on in the process design of a plant, the physical properties needs are a lot more modest than they are at the point where you're trying to get down to nitty gritty detailed design. So, in some senses I'm not suggesting that you would ever design anything, particularly something that would put people's safety at risk, based solely on theoretically computed properties." However, Cummings was aware of one instance in which a chemical plant was designed based on computed properties from quantum chemistry, but did not know if that plant was ever built.

Sandler returned to the safety issue. "When one thinks about safety, it's really reaction kinetics that gets very important. There we could have a significant time-scale problem. I'm just wondering what you can tell us from your study or elsewhere about studying reaction kinetics by simulation." Sandler also asked about calculations where safety is an issue that involve computational fluid dynamics, with mixing and diffusion problems.

Cummings said he knew a limited amount about this issue, and referred those interested to a recently published ACS symposium.[6] "I'm not somebody who does quantum chemistry calculations. I agree with you that an important safety issue is runaway reactions and how you model those in a way that enables you to design systems in a safe way and to operate them in a safe way. All I can say is the reaction we got from people in industry was that they are using these computed properties, but they are using them for things like enthalpies of reaction, things like that, which then have to be put into a process scale model along with CFD and reaction kinetics models and so on. In some sense I don't see a real big role for molecular simulation in that process, just as I was answering John [Kincaid] at the back there, I'm not sure I see an important role for molecular simulation in this area, except, as you mentioned, in diffusivity and related transport properties. It seems to me that most of the issue there is the reaction-rate constants."

Mountain offered a comment on simulations and error bars. "One of the things I think that needs to be worried about in terms of the simulation is getting a feel for how reliable the predictions are and what are the uncertainties, and if we can provide people with an indication of what they can expect in the way of error bars in terms of how the computational values will

---

[6] **Irikura, K. K. and D. J. Frurip (eds., 1998).** *Computational Thermochemistry: Prediction and Estimation of Molecular Thermodynamics.* ACS Symposium Series 677 (Washington, DC: American Chemical Society).

deviate from the experimental values. Then that becomes a tool that can be used with some confidence. That's certainly the case in the quantum chemistry business and I suspect it's going to be the case for some of the other simulation things we do, too."

Cummings agreed. "The whole point about physical properties, whether they're derived from experiment, theory/correlation, or from quantum chemistry calculation or molecular simulation, is they become useful to an engineer only when they are accompanied by realistic estimate of their accuracy. The important aspect from a process design point of view is you have to be able to propagate the errors to see the sensitivity of the cost of your design to the properties that went into it. Now, this is a big challenge for molecular simulation. How do you provide a good measure of the accuracy of a force field, because the accuracy inherent in the calculations themselves is relatively easy to estimate. You're talking about the error inherent in the force field and there we are in a sense using the same kinds of techniques that people are using for engineering correlations themselves. Basically, you have to compare to a lot of experimental data and get a feel for, within a homologous group for example, how well a force field performs. But that's something that in our community we have not done very much, if any, of in the past and something, which in the future will be very important if we're going to see, for example, industrially important and industrially respected physical properties databases take properties that are derived computationally and put them in there."

Hanley posed a question that a "neophyte" might ask. "I have a system which consists of ethylene dioxide, chicken fat, sulfur dioxide and a few bits of granite floating in it. What's the pressure, if I change the temperature by so many degrees? Now can computer simulation do a better job than actually putting that lot into a PVT cell and measuring it?" Cummings joked, echoing a humorous statement about experimentalists made by Bill Wakeham in his plenary lecture: "We'll do the pures first." Hanley countered: "You see. You're just like the rest of us at heart. I think the experimentalists win."

Pellegrino asked a question of TeGrotenhuis. "The mundane things are probably not going to be done to a great extent in micro-electromechanical and microscale devices. What do you see in the future? What are some of the things that don't come to mind immediately that will be targeted by microscale process devices? I know fuel cells for cell phones, that's already starting up. Motorola has a program in that and we have a lot of dynamics going on at those scales. What are some other things like that and the complex fluids and solid type of behavior that we are going to need properties for?" TeGrotenhuis replied: "I think where we are seeing the most activity right now in microtechnology is where size and weight matter. That's clear from both the automotive applications and the NASA interest as well. And if you think in terms of chemical processing, I think most people advocate that a just-in-time chemical manufacturing is where this makes most sense, where you have a highly hazardous intermediate or chemical that you don't want to have to store or transport. So the DuPonts of the world are looking to make chemicals in a small quantity where it's needed as opposed to big, large bulk manufacturing facilities. In terms of what complex fluids, I think in my mind where we could benefit the most is really in surface chemistry and surface properties where, to generalize a little bit, at this point we still tend to think like traditional chemical engineers and we want to try to do things the same way we've always done them at large scale at the microscale. I personally believe that we need to figure out ways of doing things at the microscale to take advantage of the things that happen at the microscale, surfaces and taking advantage of surface properties and surface tension type effects and use those as opposed to trying to work around them. And I think that's a real challenge in this whole area."

Friend asked specifically about new property needs. "Is there a specific set of fluid-surface interactions which ought to be studied? Whether stainless steel or silicon with a particular set of fluids?" TeGrotenhuis replied that the primary current interest was aqueous interactions, and as for surfaces, "we typically work in metals. Most of the rest of the microtechnology world tends to work in glass and silicon although there is more and more effort in moving to plastics as cheaper materials." Cummings noted that "when engineers came into physical properties measurements they brought metals with them," a point made in Wakeham's plenary lecture.

TeGrotenhuis added that "we tended to get in this whole area in a big way at high temperature where we had to go to stainless steel and even other more exotic alloys to go to higher and higher temperatures."

To conclude the first discussion period, Siepmann made some additional points in support of simulation. "If I ask people who teach thermodynamics, there's something called a lower critical solution temperature. The way we explain that to our students is just waving hands. We really don't have a clue what happens on a molecular level. We talk about azeotropes as being very important. Thermophysics, the experiment, can measure it, but they can't really give us a hint why there is an azeotrope. What is really the molecular structure at the azeotrope? And I would argue that simulation is the only hope, really, to trying to understand these things. We might not be as accurate as the experiment but maybe we can lead to a greater understanding and that can really help in the design of systems even if the data are not as accurate. For example you might really understand what is going on and you might really design solutions like new lubricants, if you really know on a molecular level how they work." Cummings agreed that new understanding was a major outcome of molecular simulation studies. At this point, the first discussion period was ended, and presentations by the panelists resumed.

5. Nuclear Waste Cleanup

James Poppiti

The next panelist to speak was Dr. James A. Poppiti, Team Leader of In-Tank Characterization, Department of Energy, Office of River Protection. Poppiti presented the point of view of a self-described "practical, hands-on, operations guy" involved with an urgent practical problem, the cleanup of radioactive waste at the Hanford, Washington site, and with experience in the inadequacies of models used in his work.

Poppiti described the Hanford site in some detail. The entire site (Fig. 25) covers 560 square miles (1450 km^2). Most of his work was in the 200 Areas (Fig. 26), consisting of a west area and an east area separated by five miles (8 km). He provided some historical background. "That's along the rivers where the uranium was turned through nuclear reactions into plutonium. The fuel was shipped into the 200 Area and the chemical processing extracted the plutonium for weapons. T plant is where the plutonium was first extracted, and that plutonium went into the weapon that was detonated at Alamogordo (New Mexico), and then subsequently plutonium was manufactured for the bomb that was dropped on Nagasaki."

A detailed diagram of the 200 Area was shown (Fig. 27). "There are 177 tanks. Actually, there's more than that if you count the miscellaneous tanks, but we're talking just the big ones. Twenty-eight of them are double-shell tanks; 149 are single-shell tanks; 67 of

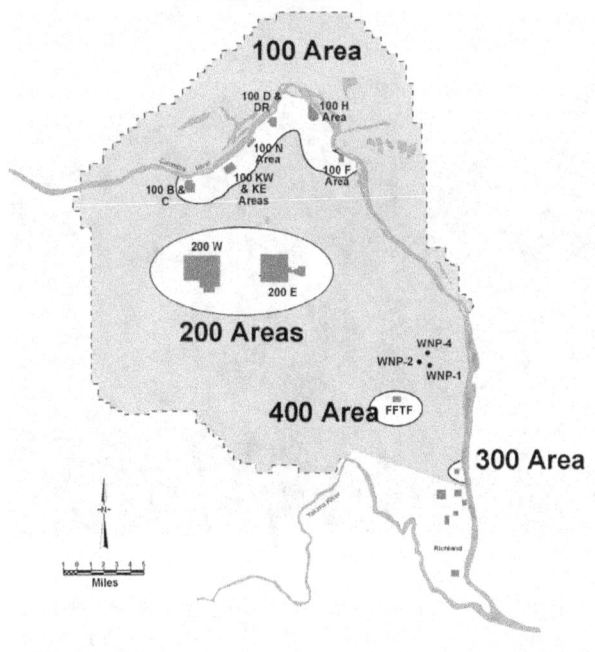

Figure 26

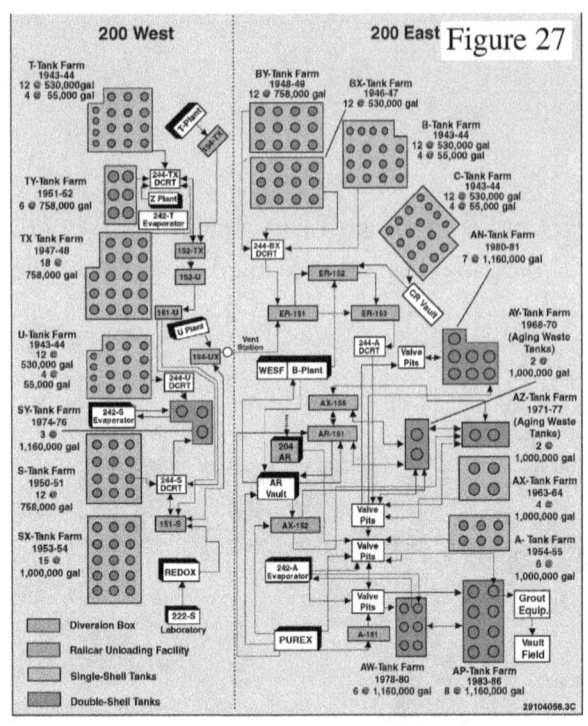

Figure 27

the single-shell tanks have leaked. The estimate is that about one million gallons (3.8×10^3 m^3); it's probably more than that, and a million curies. That's just on the leaks. All these single-shell tanks have been shut down; we're not using them anymore for obvious reasons. The double-shell tanks are still active, all 28. There are 3 tanks on the west side. We collect fluids and pump it across this transfer line to this workstation." Poppiti then showed the location of valve pits and transfer lines currently being installed.

He described the immense magnitude of the transfer problem. "The first transfer we do will be tank AT101. When we make the first transfer out of that tank, we will move in one transfer more waste than Oak Ridge has moved in 15 years. Next tank up is AZ101. When we make that first transfer, we'll move three times more curies than there is in West Valley."[7]

Poppiti showed some photographs of the tanks, typically 75 feet (23 meters) across, at the Hanford site (Figs. 28, 29). "This is a double-shell tank. There's obviously an inner shell and an outer shell. There's about 2 feet (70 centimeters) annular space. The annular space is ventilated to keep the tank cool. Obviously, we have leak detection systems. We have continuous air monitors (CAMs), with HEPA filters on the annular space. We don't have CAM alarms on the inner space because obviously you're going to pick up some radioactive material. The bottom line is, if the tank leaks, you're going to pick it up in the CAMs first. Just another quick picture – single-shell tanks (Fig. 30). This is circa about 1943 when the single-shell tank farms were under

Figure 28

---

[7] The West Valley Demonstration Project is a site 48 kilometers south of Buffalo, New York, for the cleanup of waste from a former reprocessing plant for fuel from a civilian nuclear power plant. For further information, see http://www.em.doe.gov/benr96/wvdp.html.

construction. These are smaller tanks; these are only about half a million gallons each. They're single-shell with a concrete dome."

Figure 29

Poppiti described the sampling processes. It was possible to sample only in certain specific places, because the environment was high-radiation, and only one-inch plugs could be taken. Still, the group knew quite a bit about the waste. Liquids from the tanks were sent to evaporators.

Some tritium and technetium, which is fairly mobile and moves as an anion, ended up in the ground water. "These are salts that come from the evaporators (Fig. 31). You can see from varying colors that it's mixed with all sorts of other stuff. These are high pH solutions. The material that's left over when you dissolve the fuel, you get aluminum oxide from the aluminum cladding. Some zirconium cladding was used, but mostly this is sodium nitrate, sodium nitrite, of course, sodium hydroxide, water, some aluminum oxides, iron, etc. This is what the sampling looks like (Fig. 32). These are all contaminated areas. People are in their anti-C's (anti-contamination suits). Another look in the laboratory. These samples (Fig. 33) were taken from a hot cell, but you can see that there's salt, liquid, sludges, you name it."

Poppiti showed photographs and diagrams of the pumps (Fig. 34). They are five stories tall, and two go into each tank, with two 300-horsepower motors per pump, and a smaller motor that rotates the

Figure 30

Figure 31

Figure 32

assembly inside the tank. The intake is at the bottom.

He then gave his first example of a bad experience with modeling. A modeler said that a tank being drained was generating about 100,000 Btus per hour (29 kW). "We took samples out. The samples said we could only account for about 30 to 40 thousand Btus per hour (8.8 to 11.7 kW) of heat generation out of the tank. So we didn't know what to do, because within the transfer we needed to know what the heat load was going to be in all the tanks. So we scratched our head and we said, we plan for the high number, obviously. As it turned out, the model was wrong. The Btu content was really closer to 30 to 40 thousand. The problem came in from—I'm not a modeler, but they didn't seem to take into account—the ventilation system. This particular tank was under active ventilation." Poppiti showed temperature history diagrams of the tank being drained (Fig. 35) and the tank into which the transfer was made (Fig. 36).

As a second example of an inadequate model, he described the problems of obtaining a permit for particulate emission into the atmosphere during the transfer. "We talked to the modelers again. We said we have some organic material in the tank. We needed to get an air permit when we were doing this from the state. We said 'how much do you think is going to come out?' They said, '20 parts per million'. We said 'OK'. We got a permit for 50 parts per million. We started the thing up and we hit 500 parts per million of total organic carbon in the first hour doing the transfer. So, whatever model they used didn't work very well. As it turned out, basically from the chemistry, what was going on was there was an organic chemical in there that was undergoing radiolysis. It formed a much smaller fragment which wasn't accounted for. Of course, when we started the pumps up and with the ventilation system, then the small fragments came out as heptene—it came out as a breakdown product from diethyl hexyl phosphate which formed a heptene. Heptenes came out. Anyway, we had to shut the system down and had to go back and beg for a permit to increase our

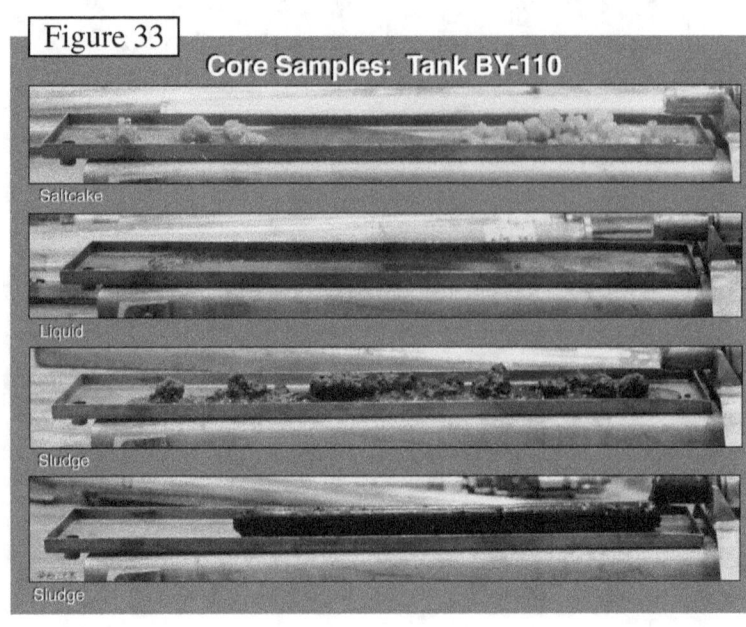

Figure 33

emissions. And then through operational means, we got the emissions level down".

Poppiti's last figure showed a breakdown of the contents of the wastes. "If you look at the liquids (Fig. 37), this is a typical double-shell tank. NA means sodium; it doesn't mean not applicable. You can account for about 98 % of everything just with that there, sodium nitrate, nitrite, hydroxyl-bound hydroxyl-free. The lab breaks it out that way. I'm not sure they can really measure it. Total organic carbon, aluminum, a little bit of carbonate and so forth, and that's in the liquid. In the solid, it's almost the same, a little different distribution (Fig. 37)."

Poppiti summarized his presentation. "The point that I'm trying to make is that I deal with the practical side of it. These are things that I have to deal with every day. We're putting a system in place to move these fluids. They're complicated. My experience with models has been pretty bad."

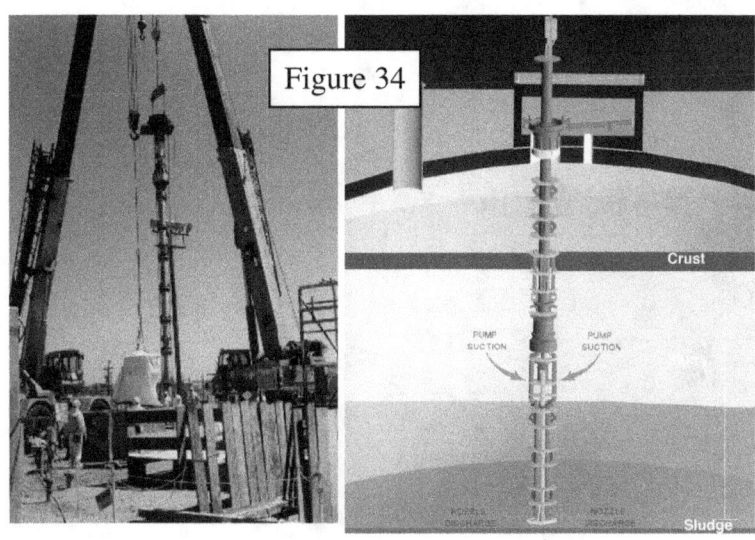

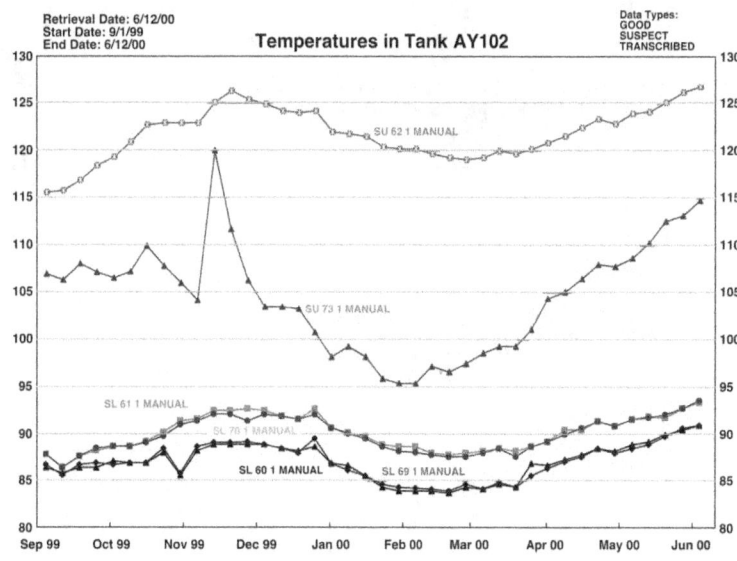

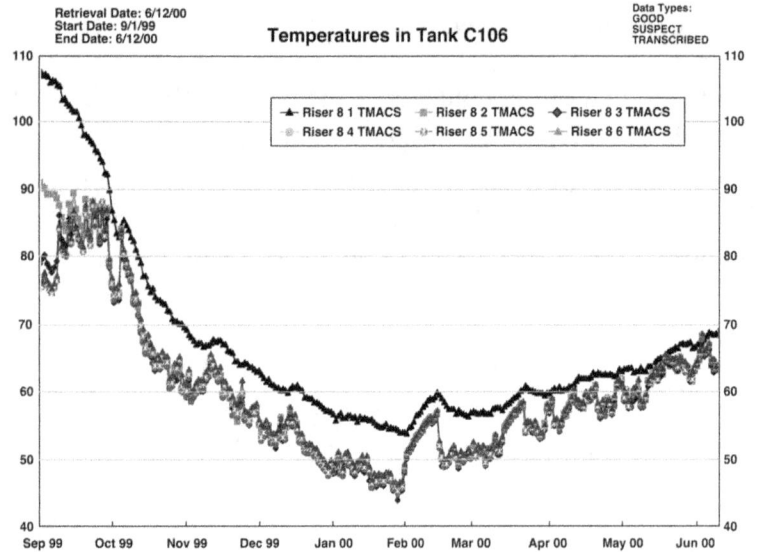

Figure 36

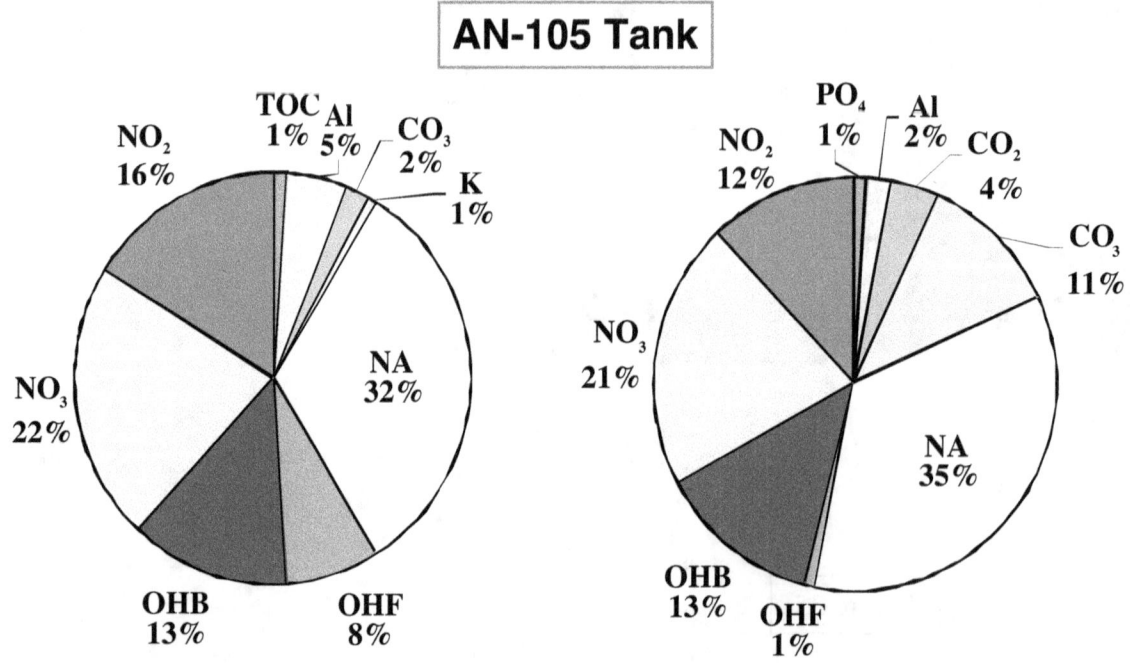

Figure 37

6. Electrolytes

Andrzej Anderko

Dr. Andrzej Anderko, Vice President, Properties of Fluids and Materials at OLI Systems, Inc. (Morris Plains, New Jersey) presented the next talk, titled "Fluid Properties for New Technologies: Electrolyte Systems." In his first two overheads (Figs. 38, 39), Anderko outlined his presentation about technologies of interest involving electrolyte solutions and the connections with thermophysical properties.

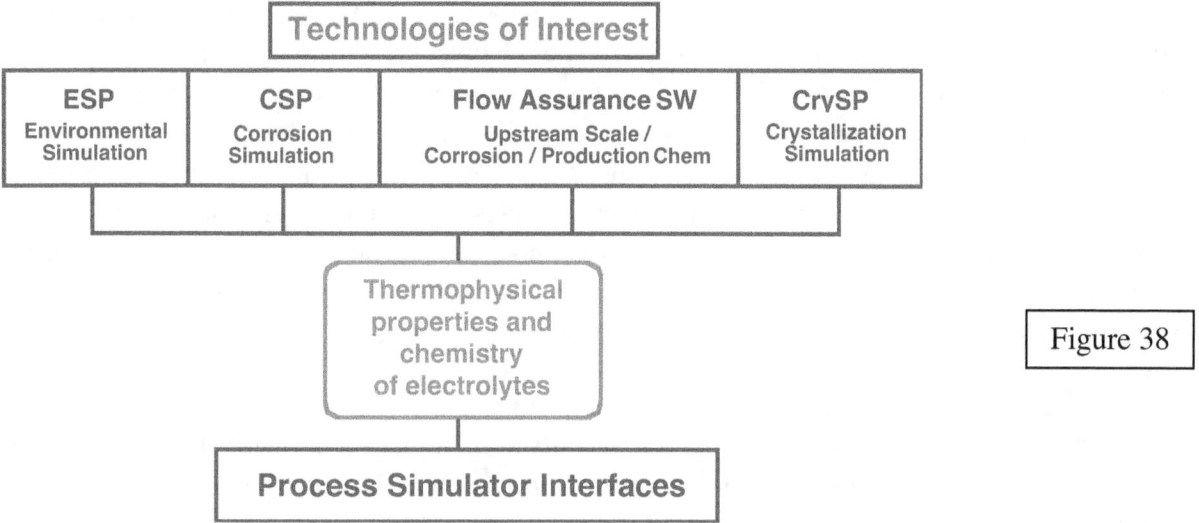

Figure 38

Dr. Anderko identified the technologies of interest. "First is environmental simulation. This is like process simulation but with emphasis on environmental applications. Second, corrosion simulation, in an attempt to come up with computational models that can approximate the behavior of real metals and real corrosive systems. Then, what we call the flow assurance software. This is essentially a combination of software for predicting scaling and corrosion in upstream oil and gas installations, and finally, crystallization simulations."

"What all these technologies have in common is a framework for computing the thermophysical properties and chemistry of electrolyte systems, because this is the fundamental underlying property on which we can build higher-level software. We also have interfaces to process simulation, because electrolyte properties are very common in industrial systems, so we want to interface it with the generally applicable process simulation tools."

Anderko listed the important technological applications in his presentation. "First, I am going to talk about the importance of thermophysical properties for a simulation of corrosion, so how we move from the electrolyte thermodynamic and transport properties to corrosion simulations. Then, I will mention fluid properties for supercritical waste oxidation or in general, supercritical-water oxidation processes. Then, I will try to talk about electrolyte properties for design-

Technology highlights

- From electrolyte thermodynamics to corrosion simulation: Importance of the properties of electrolyte systems
- Fluid properties for supercritical water oxidation
- Electrolyte properties for designing syntheses of inorganic materials
- Soil/aqueous systems: Combined effects of adsorption, solubility, phase splitting, etc.
- Database needs

Figure 39

ing synthesis of inorganic materials. And also, I will touch upon properties of systems containing soil and aqueous solutions, which involve combined effects of adsorption, solubility, phase splitting, etc. Finally, I will mention database needs."

Anderko described the prediction of corrosion, and the role of thermophysical properties therein (Fig. 40). "Corrosion goes way beyond just thermodynamics and transport properties. However, thermodynamics gives us the foundation on which we can build models for corrosion. The reason for that is, from thermodynamics, we can figure out what species exist in the system and which species may be active as corrosive agents. Well, pH is the simplest example, but it's not really limited to pH. It also involves various complexing agents, various species that interact with passive films and metals. Also, computation of transport properties is important, because the diffusivity and viscosity of solutions are necessary to compute the mass transfer effects, especially when corrosion is limited by mass transfer of species to or from the corroding interface."

As an example, Anderko showed a diagram of the corrosivity of three important acids (Fig. 41). "Just to show you an example: this is a really simple picture, probably the conceptually simplest situation we deal with in the area of corrosion, which is the corrosion of carbon steel in mineral acids. These are corrosion rates as a function of normality of acids, and you can see that, as expected, sulfuric acid is much more corrosive than phosphoric acid, and the reason for that is, of course, the different activity of the proton."

Anderko mentioned that thermophysical properties are just the foundation. Prediction of corrosion rates requires understanding surface electrochemical processes and the growth and breakdown of passive films. With sulfuric acid and phosphoric acid, the essential mechanisms are the same, the difference being in the speciation.

Prediction of corrosion: Role of thermophysical properties

Figure 40

- Thermodynamic equilibrium calculations
  - What species exist in the system?
  - What are the activities of corrosive species?
- Computation of transport properties
  - Diffusivity and viscosity are necessary to model processes related to the mass transfer of species to and from a corroding interface
- Thermodynamic and transport properties are used as input to electrochemical kinetics models

An interesting example of the role of speciation is hydrogen fluoride (Fig. 41). "When it's anhydrous, it's not really corrosive. When hydrogen fluoride is mixed with water, it becomes very corrosive. So, there is this transition between non-corrosive and corrosive behavior, and that is due to speciation. So, the role of the ther-

Prediction of corrosion: Role of thermodynamic properties

- *Example 1:*
  Corrosivity of acids depends on the activity of protons

- *Example 2:*
  - Anhydrous HF is not corrosive
  - Aqueous HF is very corrosive
  - Transition between the two regimes depends on thermodynamic speciation

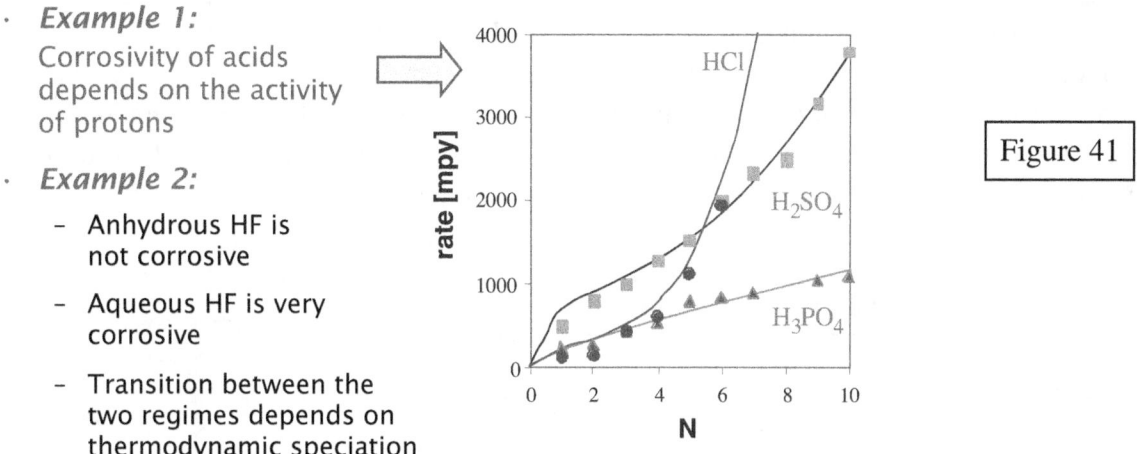

Figure 41

modynamic prediction method is to predict, for example, how many drops of water can be tolerated, and how many drops will cause the system to become corrosive."

Anderko next discussed the problem of supercritical-water oxidation and the phase diagram of aqueous sodium chloride solutions (Fig. 42). "Here, probably the most common problem is the precipitation of salt, that limits the operation envelope in which the equipment can operate, because the salt may clog the lines. Other properties are also important, like vapor-liquid equilibrium splitting, densities, and enthalpies for designing the process. So, we've been working on developing models for predicting the thermodynamic properties of such systems. And, this is just an example for the prototype system, sodium chloride and water, in which we can predict vapor-liquid equilibrium over pretty large ranges of temperatures and pressures."

Properties for supercritical water oxidation

- *Problem:*
  Precipitation of salts limits the operation envelope

- Other properties (VLE, densities, enthalpies) are also needed

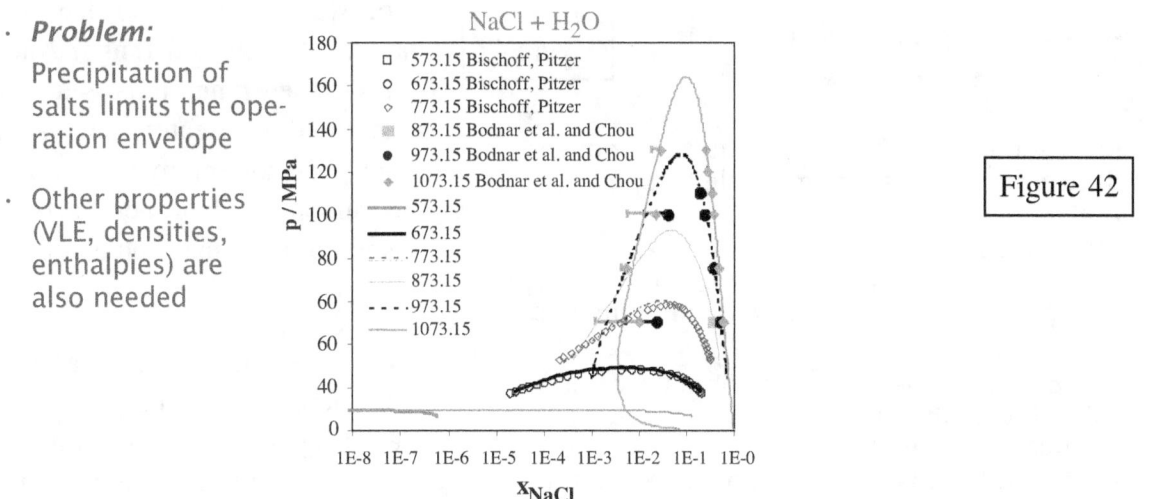

Figure 42

Anderko summarized the methodology for supercritical-water oxidation. The first requirement is the computation by thermodynamics of vapor-liquid and solid-liquid equilibria. "And, then, it is necessary to go into kinetic modeling to handle the oxidation process. The current status is that an accurate equation of state is available, but it has been parameterized for a limited number of systems. So we do have some limitations. One of the limitations is the lack of phase equilibrium data for many systems that are of practical interest. Another limitation is that the behavior of multi-component systems is poorly known, and there is an acute lack of information on transport properties in the presence of salt in such systems."

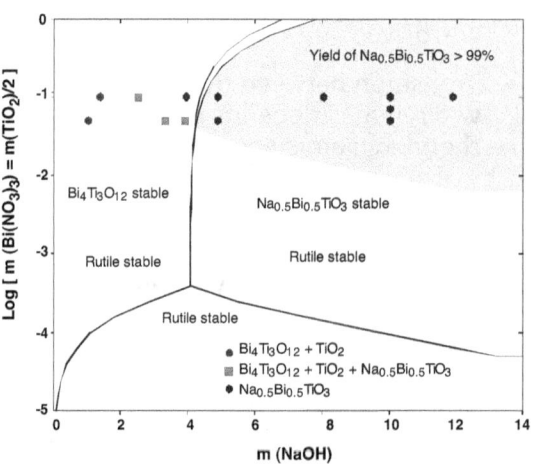

Figure 43

Properties for designing syntheses of advanced inorganic materials

- *Example:*
  Hydrothermal synthesis of ceramics: precipitation of multicomponent oxides from complex aqueous systems
- *Problem:*
  How to optimize the conditions of the synthesis?

Another example cited by Anderko is the optimization of synthesis processes for the production of inorganic materials, in particular ceramic materials. "One very interesting route for synthesizing piezoelectric or ferroelectric materials is through hydrothermal synthesis from simple inorganic precursors. This is an example of what we call a yield diagram (Fig. 43), which is like a map that shows you what concentration of the precursor material you want to select and what temperature and pressure you want to select, so that you are in this shaded range. And, this shaded range tells you that the yield is greater than, say, 99 %. So, thermodynamics and thermophysical properties help us to optimize the synthesis condition."

Figure 44

Properties for designing syntheses of advanced inorganic materials

- Methodology:
  - Equilibrium computations for multicomponent electrolyte systems with multiple competing solid phases
- Current status:
  - Syntheses of several piezoelectric ceramics have been successfully optimized and implemented
- Limitations:
  - Lack of thermochemical data for many complex solids (e.g. multicomponent oxides) and secondarily, aqueous species
  - Insufficiently advanced methods for estimating thermochemical properties

Anderko summarized the methodology for this problem (Fig. 44). "It is the equilibrium computations for multicomponent electrolyte systems

Soil/Aqueous Systems

Figure 45

- Methodology:
    - Adsorption phenomena (ion exchange, molecular adsorption, surface complexation)
    - Phase equilibria in the bulk (e.g., partitioning between aqueous and NAPL phases)
    - Solubility and speciation effects in the aqueous phase
- Current status:
    - A comprehensive model has been developed and verified against experimental data
- Limitations:
    - Parameters cannot always be determined because of insufficient characterization of systems of interest
    - Kinetic effects are not well characterized

with multiple competing solid phases. And, this is the key: the multiple competing solid phases. Currently, synthesis of several piezoelectric ceramics have been successfully optimized and implemented. So, this really works; but, there are limitations, especially with respect to a lack of thermochemical data for many solid species, which have to be estimated."

The final technology Anderko discussed was the important problem of modeling soil and aqueous systems (Fig. 45). "Here, we are dealing with many combined phenomena, like adsorption phenomena, ion exchange, molecular adsorption, surface complexation. We are dealing with phase equilibria in the bulk, like liquid-liquid splitting between aqueous phases and non-aqueous petroleum liquid. So, it has been possible to develop a comprehensive model. However, the parameters of such a model are really difficult to estimate, because the systems of interest haven't been really characterized with sufficient detail. And also, kinetic effects are not well characterized, and they are important in subsurface systems."

Anderko's final viewgraph outlined the needs for databases (Fig. 46), and in closing he summarized his presentation. "We are trying to leverage thermodynamic properties and thermophysical properties of electrolyte systems to a variety of applications. And, from all those examples, it is very clear that they are very important."

Database needs for mixtures

- Comprehensive, computerized collections exist for nonelectrolyte systems (DECHEMA, TRC)
- Several printed data sources exist for electrolyte systems at "normal" conditions. However,
    - No computerized collection is available
    - Many valuable sources have not been updated in decades
- No database exists for high-temperature and supercritical electrolyte systems

Figure 46

7. Second Discussion Period

The second discussion period followed Anderko's presentation. Laesecke began by noting that there were many attendees from outside the United States who might not realize the magnitude of the U.S. nuclear waste cleanup problem. He was recently at the Trinity site and emphasized that he understood the scope of the environmental problem only through his visit. "We have Rocky Flats just eight miles south of us [south of Boulder, Colorado] which is a shutdown nuclear-weapons plant, and you should understand that Hanford was the site where plutonium was prepared for the first nuclear bomb which was exploded in New Mexico at the so-called Trinity Site July 16th, 1945. The Department of Energy has shut down all the nuclear weapons production." He mentioned the Hanford site in Washington, Rocky Flats in Colorado, Oak Ridge National Laboratory in Tennessee, and the Savannah River site in South Carolina. "This whole chain of nuclear weapons plants is shut down and the effort is now to clean the contamination," much of which is in stored tanks. "That's an environmental problem of unthinkable magnitude." Laesecke noted that Poppiti had indicated "that in one transfer from a tank [at Hanford site], they transfer three times more than Oak Ridge ever produced in 15 years."

Poppiti offered some elaboration. "The first transfer [at Hanford] had more volume than Oak Ridge transferred in 15 years, and one transfer will transfer 3 times more curies than at the whole reprocessing plant at West Valley, New York. There are 54 million gallons of waste in those tanks. There's the most waste there in the entire complex, and we're the only site that has no treatment capability."

Laesecke posed a question on priorities, in which he compared the technical challenges explored by TeGrotenhuis, such as the NASA program to produce propellant on Mars for the return of a space vehicle to Earth, with the vast environmental challenges facing the Hanford operators.

Poppiti further described his problems with modelers. "The two stories that I told about modeling are just two that I picked out. I have a whole bunch more. What I'd really like to have is some model that I had some confidence in. Every time that I get predictions, more often they are wrong rather than right and I don't mean wrong by 5 % or 10 % or anything like that, I mean really wrong enough that you have to shut the process down." He said the resulting shutdowns were intolerable: they lead to additional sampling, delays, etc. "We've got compliance agreements, compliance orders among other things to do certain things by certain times. As an end user, I'm very frustrated."

Hanley offered a follow-up comment. "I think this is the kind of thing that the thermophysical property community should be listening to. These problems are of such gigantic scale that serious rethinking is needed. I guess if you were writing a proposal to DoE, you would try to come up with a model that would work." Poppiti interjected, "that'd be refreshing." Hanley continued, "but the problem is that a lot of the people in this room, no disrespect, don't even know how to start. It's a chicken and egg thing. We ask questions of people, such as yourself, and we don't even know really what questions to ask and you don't know how to prompt us. I'll put it another way. If you wanted data, what data do you need to solve your problems?"

In reply, Poppiti said, "I can present it pretty simply. We were making a transfer from one tank to another. You have an active ventilation system with a hundred cubic feet a minute, or whatever it was running at. I don't remember the exact number. We took samples of the material. So there was plenty of material. People could analyze it and ask for anything you want. And then we said, 'okay, this is the real world, guys.' You have to go get a permit before you start the operation. We went to the modelers and said, 'what should our emissions permit be?'"

Cummings then made a point about modeling. "When we talk about modelers in this context, we're talking about process level modelers, not molecular or quantum chemistry type modelers. I think what you're seeing is the inadequacy of the physical properties for these process level models and possibly of the process level models themselves, rather than something deeper down the chain." Poppiti replied, "I'm not sure who I need to talk to, to get an answer. All I know is, when I turn the pump on, I don't want to have to turn it off right away."

Hanley commented on his experiences with the Department of Energy (DoE). "I've had a lot to do with DoE and, frankly, there's a great deal of frustration on the side of people like me." Hanley observed that DoE does not have a transparent mechanism for informing researchers about specific applications needs, and "I don't know what they want," so in the end "no one gets anything." Cummings replied, "the DoE has part of it which is environmental management, which is the clean-up part of DoE. The part of DoE which you most likely interact with is the Office of Science, which is a funding agency. There've been efforts to try to inject more science into environmental cleanup, the Environmental Management Science Program for example. But the environmental needs are so large, so compelling, and requiring such immediate action, that these things have not really panned out the way people had hoped they would." Hanley agreed, but added, "I think the moral of the story is that if academic (I use the word academic loosely) people wish to cooperate and, particularly, to contribute, then they have to go to the sites themselves and see the problems. It's quite awe inspiring to see a huge tank which is rotting away." Cummings noted that there is a similarly large clean-up of old storage tanks underway within Oak Ridge National Laboratory.

Outcalt offered a viewpoint regarding DoE. "The environmental science program at DoE originally said that they wanted to look at basic sciences to start in the huge effort that you have to clean up Hanford, and I think they've moved away from the basic sciences. I can appreciate the need to clean these things up as fast as possible, but in moving away from the basic science I think that you are missing a lot of the basic measurements and the needs that you have to accurately characterize the systems that you are working with. That is being reflected in the fact that the models are not coming out with good predictions. It is possible that there's not accurate data that's been taken to model the systems as well as you need them modeled." In response to a question from Hanley, Outcalt said that, as well as problems with the model or the input data, the problem could be lack of data.

Anderko offered an additional comment about modeling. "Such systems are so complex that any process-level model has to be calibrated against all the building bricks. Little building bricks like binary systems, or like absorption characteristics if something gets into the soil. So it's a huge experimental program, and only after such a program is complete can you hope to have an accurate process-level model." Poppiti responded by showing again some of his slides, and strongly reiterating that he had a well-defined problem for which appropriate modeling was a necessity.

Joback offered a comment on modeling difficulties and the education system. "There's basically a gap between real world applications and really what properties we're looking for. For example, in this case it's not so much that you didn't have the right densities or viscosities. It's that somebody didn't realize that there might be another reaction going on. You can have great models for viscosity, but when you make a real lubricant nobody cares about viscosity. It's all surface effects and such. When you're going to make a new bumper for a car, as far as dealing with some of the fundamental properties, what do you need to make a good bumper for a car? And the question is where is that expertise? Where is that connection between the fundamental properties and what you actually need to make a real product today? And it's that product design knowledge that used to be known, that I'm not sure it's being taught anywhere anymore. We deal with a lot of the fundamental properties where it relates back to the molecular structure, but when somebody says, 'I need a new paint,' 'I need a new lubricant,' or 'I need to figure out how I'm going to pump this stuff around,' where's the expertise that the engineers used to have to say, this is what the properties are, and this is what you have to measure?"

At this juncture, Hanley called for an intermission. During the intermission, reporters from the organizing committee compiled a set of "bullets." These were the primary themes of the discussion, and Hanley used these as input for his impromptu summary (see Section 11) of the Forum.

8. New Generation of Power Plants

Thomas O'Brien

The first panelist to speak following the intermission was Dr. Thomas J. O'Brien of the Department of Energy, National Energy Technology Laboratory (NETL), Morgantown (West Virginia). O'Brien first showed an organizational chart of the Department of Energy and the role of NETL (Fig. 47), which has a sister site in Pittsburgh (Pennsylvania). The laboratory is not involved in classified research. "We don't have any secrets. We have nothing to lose, nothing to hide." All fossil energy programs come through that laboratory.

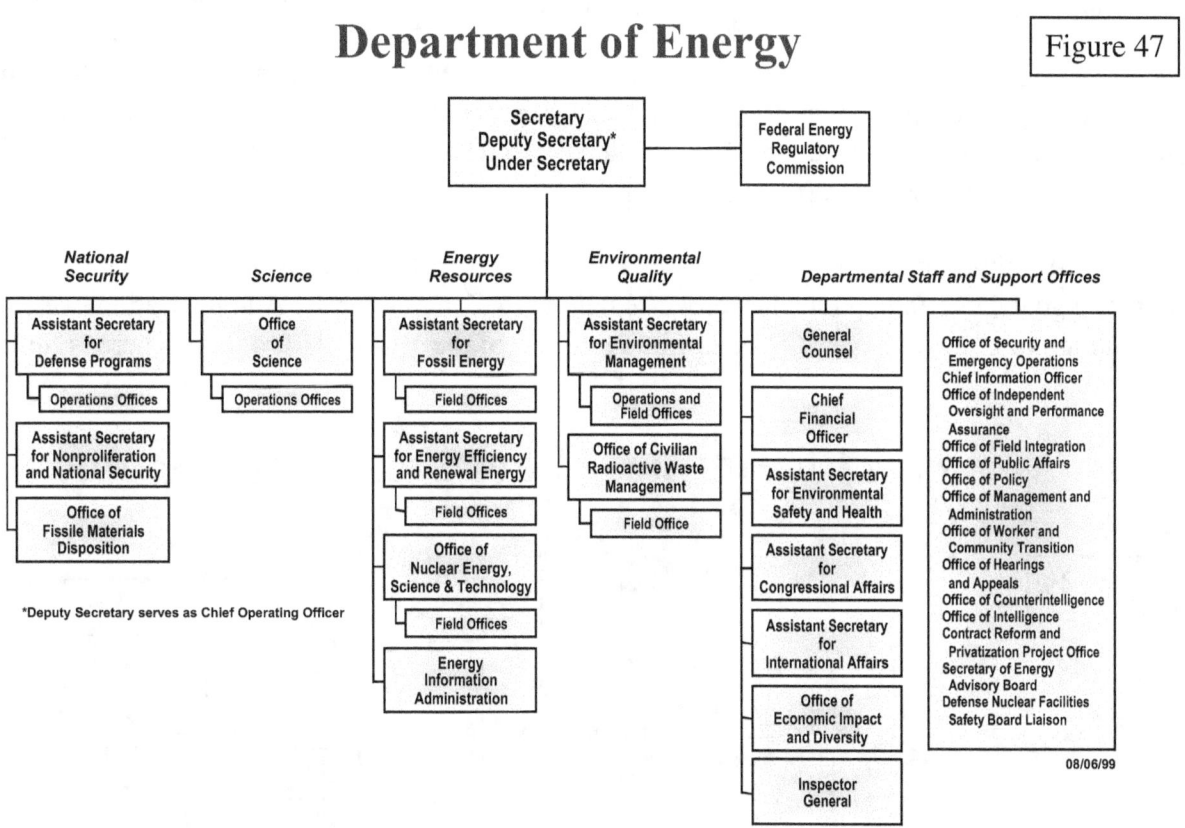

Figure 47

He introduced the main subject of his talk, the Vision 21 program. "We have many ongoing programs in fossil energy—fluidized-bed combustion of coal, turbine combustion of gas, fuel-cell program, solid-oxide fuel program specifically, but the program that we're cranking up for the next 15 years is something that we're calling the Vision 21 program. But, the idea of Vision 21 is basically to strongly crank up the efficiency, remove the environmental constraints, so that burning fossil fuel, so that coal is not a nasty word anymore. Coal is green rather than black."

Figure 48

Vision 21 Performance Targets

- **Efficiency - electricity generation**
  - 60% (HHV) coal-based; 75% (LHV) gas-based
- **Efficiency - fuels only plant**
  - 75% (LHV) fuels utilization efficiency
- **Environmental**
  - Near-zero emissions; 40-50% CO_2 reduction by efficiency improvement, 100% with sequestration
- **Costs**
  - Competitive
- **Timing**
  - Major benefits by 2005; module/plant designs by 2012/2015

He listed three premises as the basis for the program. The U. S. will need to rely on fossil fuels for transportation well into the 21st century. It is prudent to rely on a diverse mix of energy resources. Better technology can make a difference in meeting environmental needs at acceptable costs.

Vision 21 includes some very specific and ambitious goals for efficiency (Fig. 48), as well as the goal of removing all concerns about environmental degradation. "The highest efficiencies for combustion with natural gas—at the present time, the most sophisticated system is 60 %. Typically, power plants that you see sitting alongside the road are 35 % efficiency, so this is a very ambitious goal to attain; for example, this number 75 % efficiency for a gas-based system."

An important but not fully decided issue is whether to include carbon dioxide sequestration as a goal. "Depending on whether we're talking to Congress or the Administration, we either are or are not going to have CO_2 sequestration as part of the Vision 21 program. There is, in fact,

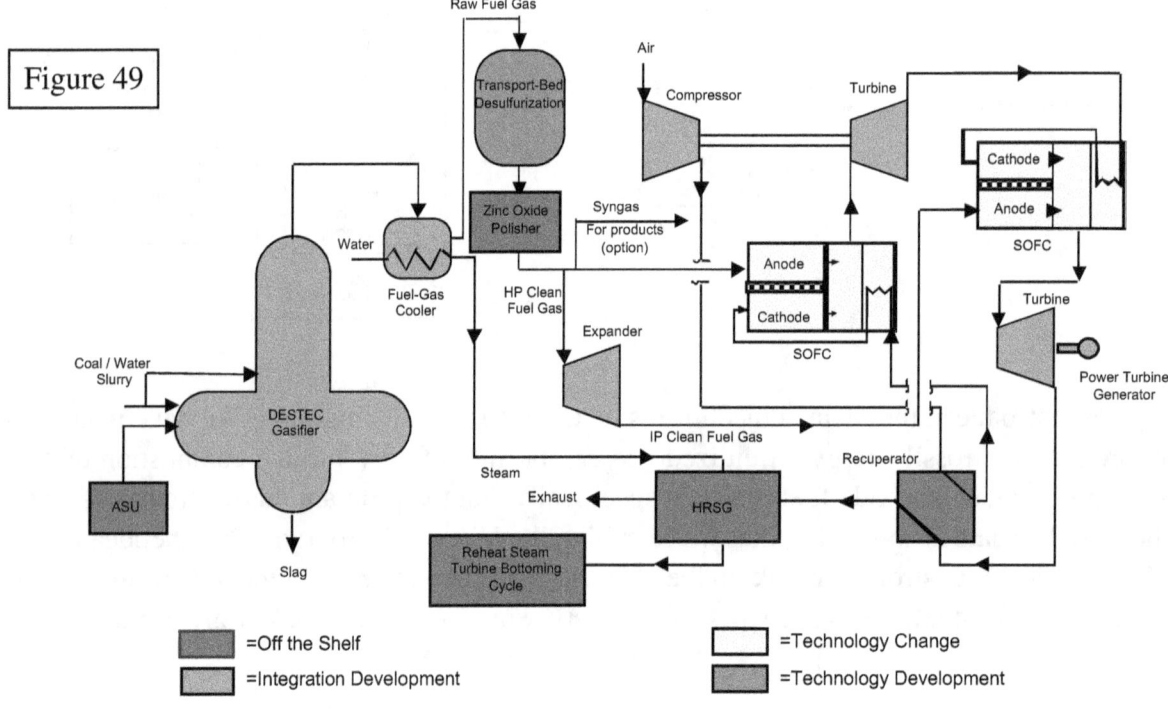

Figure 49. Vision 21 Example - Fuel Cell / Gas Turbine

a CO_2 sequestration program which is separate from the Vision 21 program in the fossil energy program, and this is what is going to be the fossil energy program over the next 15 years."

As an example of a Vision 21 product, O'Brien showed a hybrid of a fuel cell and a turbine as a schematic diagram (Fig. 49). "This is a very schematic idea of a Vision 21 plant, and I want to talk to this very schematic level. The idea is that probably we're going to have some upstream separation of air, so most of the systems will be oxygen-fired. So, that will require gas separation plants on a much larger scale than presently exist. And, there will probably be some upstream processing of the fuels. We'll be using coal. Certainly, natural gas will be very cheap and abundant for the next 15 years. But, we also want to have opportunities to use biofuels so we can swing into the CO_2 sequestration issues."

He described the processing of the fuels to be used. "We will have a whole range of fuels that we want to be able to process. And we'll probably process these in some way other than combustion. So, there will be some sort of gasification system that will transform these into a higher-intensity fuel. Then that will cascade through the power plant cycle. There'll probably be separating off some sort of stream which will be accessible for CO_2 sequestration and will cascade through a series of conversion devices."

Near-term Turbine/Fuel Cell Hybrid

Figure 50

- **SOFC Siemens-Westinghouse demonstration at National Fuel Cell Research Center**
- **Output rating - 250kw**
- **Efficiency- 55-60%(LHV)**
- **NERC micro-turbine**
- **1 MW systems planned**

He displayed a photograph of a prototype plant (Fig. 50). "This is an example I wanted to show of something that is being shipped today to the University of California at Irvine, I believe. This is what we would call a prototype of a Vision 21 plant. It's a hybrid of a solid-oxide fuel cell and a microturbine. You can see the efficiencies are still well below the goals of the program, although the efficiency of this device is nominally at 60 % conversion. This is the kind of thing that we envision, coupling something with a very slow time-scale solid-oxide fuel cell with a very rapid time-scale microturbine."

Figure 51 **Vision 21 Enabling Technologies**

- **Gas separation, e.g., membranes for separating oxygen from air**
- **Gas stream purification**
- **High-temperature heat exchangers**
- **Fuel-flexible gasification**
- **High-performance combustion systems**
- **Fuel-flexible turbines/engines**
- **Fuel cells**
- **Fuel and chemical synthesis reactors, and improved catalysts**
- ...

The Vision 21 program will require many new and existing technologies. O'Brien divided them into "enabling" technologies to develop all the necessary components (Fig. 51) and "supporting" or "cross-cutting" technologies (Fig. 52). Obtaining the desired efficiencies will require the traditional techniques of going to high temperatures and pressures, with materials to withstand such conditions.

O'Brien proceeded to elaborate on the last item of Fig. 52, virtual demonstration. "The idea of the virtual demonstration is that (this is the fight we've been having this afternoon) there are the theorists and the experimentalists. And all I'm trying to suggest is that there's a crossover point. Certainly, computer costs are going down. The algorithms are improving considerably, and so this says that simulations will be used more. The experiments are becoming more costly, especially at high temperature and high pressure, and at large scale, if we want to demonstrate these things."

O'Brien described virtual demonstration as a replacement for traditional demonstration plants. In his home state of West Virginia, in past years, the Federal Government provided "literally billions of dollars" for demonstration plants under the clean coal program, but "that is not going to happen again." He listed various virtual demonstration projects of other agencies as prototypes (Fig. 53).

"There is, of course, the grandfather of all demonstrations programs, the ASCI (Accelerated Strategic Computing Initiative) program in the weapons program of the Department of Energy that simulates nuclear weapons rather than the testing of nuclear weapons. NASA has the Intelligent Synthesis Environment program which is to be able to simulate before you send people into outer space. The idea is doing this model-based procurement kind of idea, to be able to develop models of complex weapons systems as you build them, aircraft carriers or advanced fighter planes, engines."

Vision 21 Supporting Technologies

Figure 52

- **High-temperature materials**
  - Alloys and ceramics with improved strength, durability, and corrosion and temperature resistance

- **Advanced controls and sensors**
  - Improved hardware and software to monitor conditions directly, detect early signs of failure, and manage complex processes

- **Environmental control technology**
  - Control NO_x, fine particulate, trace metals, manage byproducts
  - CO_2 separation

- **Advanced manufacturing and modularization**
  - Minimize design and construction costs and improve quality

- <u>**Virtual demonstration**</u>
  - Advanced computer models for simulation and visualization

O'Brien described collaborations with other Federal agencies. "In fact, we're cooperating with both NASA and the Air Force in the advanced engine development and simulation capability because they're simulating engines for aero purposes and we're simulating engines for ground-based power energy generations. So, everything in the virtual demonstration is basically everything we do on computers."

He mentioned organizations such as Aspen Technology that have done process simulation in connection with advanced visualization and CAD capabilities (Fig. 54). Control issues will be very important. "As I mentioned before, these [Vision 21 plants] are going to be coupled, probably very disparate energy conversion processes, and we may know how to control one or the other, maybe even both, if we're lucky. But, it's going to be a very difficult thing to marry these two things together, so that we will be able to handle them efficiently." He showed examples of mechanistic modeling (Fig. 55) and listed some mechanistic simulations at NETL (Fig. 56). "Where I come from, there is the idea of mechanistic modeling to be able to move very science-based models into all of these aspects of the virtual demonstrations. The summary is what I think is a cultural change that's occurring in the power industry."

Figure 53 | **Virtual Demonstration - Related Programs**

- **DOE**
    DP - Accelerated Strategic Computing Initiative (ASCI)
    SC - Scientific Discovery through Advanced Computing
- **NASA**
    Intelligent Synthesis Environment (ISE)
- **DOD**
    Model-Based Procurement
    Enterprise Model Based Lifecycle Management
- **NIST**
    Next Generation Manufacturing Systems

O'Brien noted that extensive use of computers has not been traditional at NETL. "The power industry is actually kind of an antiquated industry. Though we're in the Department of Energy, in the Fossil Energy program, computers were not a big part of the Fossil Energy program. We were not the Atomic Energy Commission. We did not have the heritage of science and computation. And, coal has been used and converted for hundreds of years. And, so it's just becoming now that in our program we're looking at simulation as a large part of the Vision 21 program. And, the idea is to have it well-balanced." In conclusion, and in answer to a central issue of Forum 2000, O'Brien quoted from the Office of Science that science is no longer based on the two legs of analysis and experimentation, but the "three-legged stool" of analysis, experimentation, and simulation.

Virtual Demo Components
Process Simulation

Figure 54

Process Optimization
Economic Evaluation
Component Sizing
Sensitivity Analysis

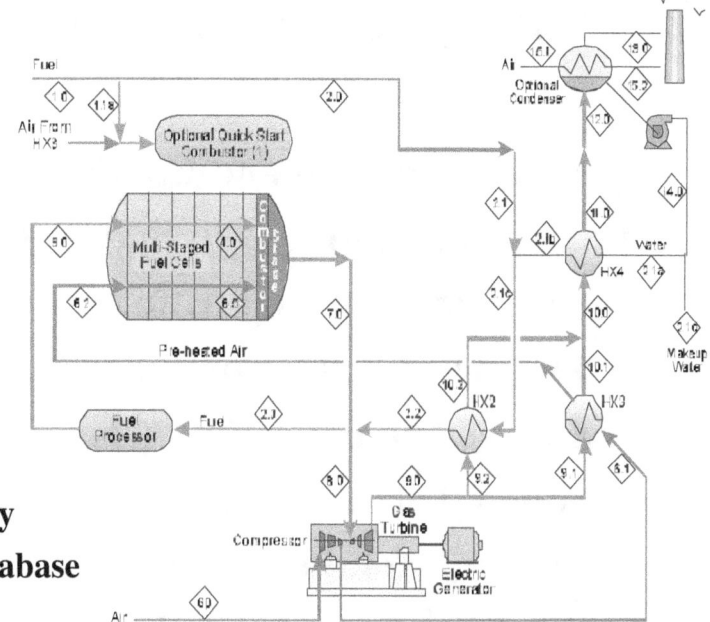

Unit Operations Library
Physical Properties Database

Virtual Demo Components
Mechanistic Modeling

Figure 55

CFD Simulations
- single/multi-phase
- heat transfer
- chemical reactions

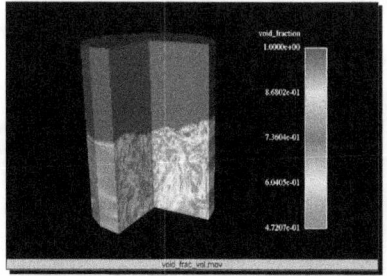

Finite Element Structural Simulations

Event Based Simulations

Material Simulations

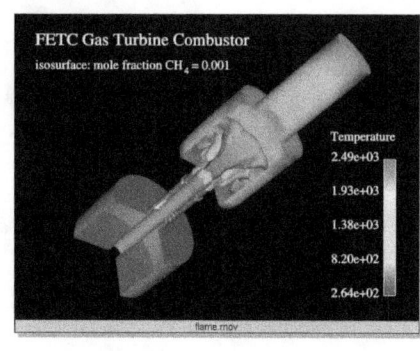

Figure 56

Examples of Mechanistic Simulations at NETL

- **Pulverized Coal and Biomass Combustor**
- **Circulating Fluidized Bed Gasifier**
- **Gas Turbine Combustion**
- **Trapped Vortex Combustor**
- **Staged Topping Combustor**
- **Solid Oxide Fuel Cell**
- **Internal Combustion Engine**
- **Chemical Industries of the Future**
- **Computer Hardware**

9. A Process Manufacturing Perspective

Paul Mathias

Dr. Paul M. Mathias, Principal Advisor with Aspen Technology (Cambridge, Massachusetts) gave the next presentation, identified in his title slide as "A Process Manufacturing Perspective." He said that what is done at Aspen Technology is essentially to simulate industrial plants, and he would review how data were used, how data have been helpful, and what the future needs are.

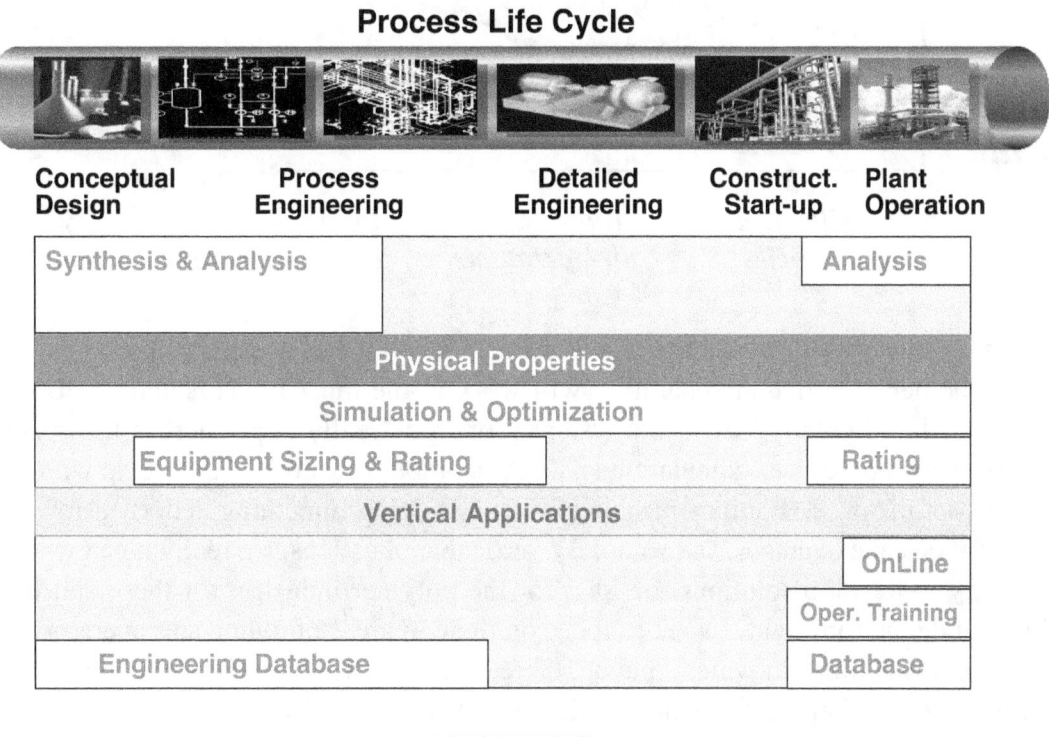

Figure 57

Mathias began with a viewgraph titled "Process Life Cycle" (Fig. 57). "This is a slide that my colleague, Suphat Watanasiri, put up in his talk a couple of days ago. And, I thought that this would be a useful way to explain the way in which we see the world of modeling, and maybe this would answer some of the questions that came up earlier. At the top left, there is the idea of conceptual design. That's a chemist coming up with a new chemical, or an engineer dreaming up a new process. And that moves on to the stages of process engineering, detail engineering, construction, startup, and plant operation. What we have provided here are the various tools that are available in the process engineering, process simulation community. And, we're trying to see where they fit in the various stages of the life cycle."

Process Design Modeling Vs. Chemical Plant Component

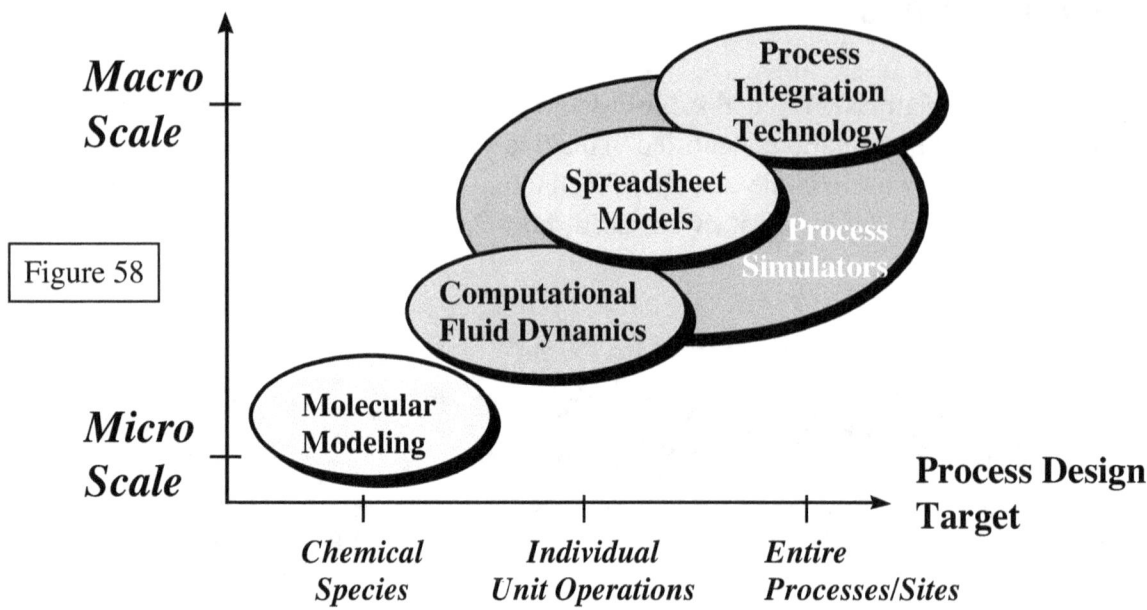

Figure 58

Mathias discussed in turn the items of Fig. 57. Synthesis and analysis means trying to understand whether a particular separation will work at the most fundamental level. Physical properties were highlighted, "because physical properties really apply across the spectrum." Following physical properties, "simulation and optimization, broadly, is coming up with a model of a plant and optimizing it. Equipment sizing is actually doing something in detail. How big do I need to make the heat exchanger, and so on. By vertical applications, we really mean what we, as a company, try to develop solutions for, say, for the polymer industry, for the metal industry. Down there on the bottom right, I want to focus on the two areas of online and operator training, because I think those areas in which, maybe, there is less participation by the community here. And, I think that those are great opportunities in the future." Also important are engineering databases.

Figure 59

"Fluid" Properties have an Excellent Success Record

- Critical data evaluation (e.g., NIST Chemistry WebBook)
- Model development (broad cooperative effort)
- Correlation development (e.g., DIPPR)
- Benefit of <u>standardization</u>

> ... but the effort has been largely restricted to petroleum, petro-chemicals and gas processing, which I will refer to as "conventional" systems

Mathias presented a schematic diagram of process design modeling vs. chemical plant component (Fig. 58). "On the *x*-axis, we have the range from chemical species to individual unit operations to the entire process. On the left would be a scale. The microscale, for example, if you heard the miniaturization talk, would be on a very tiny scale, to the macroscopic scale, where we're dealing with bulk fluid. Process simulation really has carved out a niche for itself in the area of individual unit operations through the entire process. But, it has a bit of difficulty currently going into the microscale. And I think that there would be the opportunity that we just heard about, in terms of developing the synergies between the computational fluid dynamics and process modeling. Molecular modeling, in this picture, would be the one that would, say, help the invention of materials. But, we've also talked about the production of properties that could be used in process simulation to a molecular model."

Comments on Status of Properties for "Conventional" Systems

- Education/training is a big problem
  - Reduction/disappearance of central engineering groups
  - Companies pride themselves on having generalists rather than specialists
  - Companies don't understand issues in detail
  - General software versus specific applications
- Uncertainty analysis
  - Prediction of quantitative effect on final design elusive
  - Great value in identifying weaknesses in data
- Need for transport properties

Figure 60

Comments on Status of Properties for "Conventional" Systems - Continued

- Molecular modeling for property predictions
  - Prediction of condensed-phase properties is weak
  - Just a more sophisticated data-fitting exercise?
  - Role for molecular modeling may be to understand trends
  - Hybrid models
- Weak need for new correlational models
- Software
  - Exciting opportunities for on-line applications, soft sensors, state estimation for control, etc.
  - Open architecture (Global CAPE Open) and the Internet opens the field to new players
- Database
  - Many on-going efforts

Figure 61

He then began his discussion of "conventional" areas (Fig. 59), as opposed to "unconventional" areas such as electrolytes, polymers, and solids. "These are the areas where I've seen very good contributions from areas like NIST. Especially in my previous job at Air Products, that was invaluable to me. Without the equations of state for helium, and the fluids from methane to ethane and oxygen, we wouldn't have been able to achieve the kinds of designs that we did. And, so, I really did appreciate several areas here, but particularly the critical data evaluation, which was really wonderful. In terms of the models developed, I think that's been a rather broad effort in which academia has participated, and other areas. The correlation development is an area, for example, certainly DIPPR has done it."

"But, let me highlight the last bullet item up here, and I think one of the great contributions of NIST, and indeed that's where it came from, is indeed the standardization. I know, for example, early on, when we had developed a model for helium, and gave it to a customer, they'd come up with a different number. There was always a debate about who was more accurate. So, it was great when NIST came in and said this is the model. We put our blessing on it, and this is the standard. I think this was a wonderful contribution. But, let me get down to the bottom comment there. I think this effort has been largely restricted to petroleum, petrochemicals, and gas processing."

Mathias' next viewgraph outlined the status of properties for conventional systems (Fig. 60). "I think education is a big problem that we have today, as people have become very broad. Professor [William] Wakeham gave this talk Monday, and again repeated it yesterday. And for those of you who heard it, my opinion anyway, is that a lot of those issues get down to education. For example, there's a reduction and disappearance of central engineering groups. Companies pride themselves that I'm a generalist; I don't have specialists anymore. Companies actually don't understand the processes in great detail anymore. There used to be this idea of core competencies that seems to have drifted away.

"Uncertainty analysis—this is my own opinion of it, and I'm sure people will differ—but whenever I try to do uncertainty analysis, I could never do the propagation of errors in any meaningful way. It never really made any sense to me. But, I think the great value of uncertainty analysis was that it helped identify what the weaknesses were. We knew what we had to work on. I think that was very important.

"Transport properties is a big need. I think that people, when they weigh it, always come up with phase equilibrium as being the most important, and then the energy properties, enthalpy, and so on, and transport properties, has been the stepchild. I think we need some more work on transport properties."

Mathias continued with a second viewgraph on the same general subject (Fig. 61). "Molecular modeling for properties prediction – that kind of gets into Peter Cummings' area here. I think the prediction of condensed phase properties is weak. The next question is: 'Are we just doing a sophisticated data-fitting exercise?' If we are, let's go back to fitting parameters and cubic equations of state. Although I think the role for molecular modeling is, to my mind, to understand trends and to build hybrid models.

"My own opinion is we probably don't need correlational methods. As for software, there are a couple of things, and here I'm talking about simulation software, that I wanted to highlight. One is that there is an exciting opportunity for online applications, soft sensors. For example, if somebody really wants to control molecular weight, but they can measure only viscosity, and maybe some other indicators, I think if we had the relationships from one to the other, that would be wonderful. And, that's where the physical properties community could

Issues and Needs for Electrolytes

- Education is (again!!) a big issue
- Speciation and complex formation are the key issues
- Very little new recent data
- Lesser need for new correlational models
- Important applications in environmental, corrosion, new technologies

Figure 62

Mathias turned to the topic of software and open architecture such as Global CAPE (Computer-Aided Process Engineering) Open, and the internet, which in his opinion would open up the field to new players. "It used to be that the barriers were so high that only the simulation companies could put out property models. I think now that the field is open. And actually, we welcome that, because there's only so much that we can do." He mentioned current efforts in databases as presented throughout the Fourteenth Symposium on Thermophysical Properties. As a caution, he displayed a quotation from Philip W. Anderson's article, "More is Different," Science, 1972. "The ability to reduce everything to simple fundamental laws does not imply the ability to start from those laws and reconstruct the universe."

Mathias then proceeded to "unconventional" areas, including electrolytes, polymers, polymer-electrolyte mixtures, and solids. Issues and needs for electrolytes, a subject discussed earlier by Dr. Anderko, were summarized on Fig. 62. As an example, Mathias posed a question: What is the pH of a 1M $FeCl_3$ solution? He showed a table of the many chemical reactions of $FeCl_3$ in water (Fig. 63), and the detailed equilibrium distribution of molecules and ions in such a solution (Fig. 64). "The key thing is the chemistry model, and that's something I'd like to urge groups here to work on. Given the chemistry model, one could say that the pH is 2.2, because there's this particular hydroxyl compound that appears from this particular model." More generally, he showed a diagram of pH versus molality (Fig. 65), which he said could be relatively simply generated from a simulation. He then reiterated the importance of speciation by displaying a sign he keeps on his wall which read "It's the speciation, stupid!"

Mathias continued with the issues and needs for polymers (Fig. 66). "The key thing about polymers, I think, is that first bullet item, when we do simulations, is that it's segment rather than component-based. And the second one that's important is that it's history-dependent. Let me move down to the last item, and that is the kinetic models are really the key thing.

$FeCl_3$ Aqueous Chemistry

Equilibrium	$FeCl_3$	$\leftrightarrow$ $FeCl_2^+ + Cl^-$
Equilibrium	Fe_2OH_2+4	$\leftrightarrow$ $2\,FeIII+3 + 2\,OH^-$
Equilibrium	$FeCl_2^+$	$\leftrightarrow$ $FeCl+2 + Cl^-$
Equilibrium	$FeCl_4^-$	$\leftrightarrow$ $FeCl_3 + Cl^-$
Equilibrium	$FeCl+2$	$\leftrightarrow$ $FeIII+3 + Cl^-$
Equilibrium	$FeOH^{2+}$	$\leftrightarrow$ $FeOH+2 + OH^-$
Equilibrium	$FeIIIOH_3$	$\leftrightarrow$ $FeOH_2^+ + OH^-$
Equilibrium	$FeIIOH_4^-$	$\leftrightarrow$ $FeIIIOH_3 + OH^-$
Equilibrium	$FeOH+2$	$\leftrightarrow$ $FeIII+3 + OH^-$
Equilibrium	H_2O	$\leftrightarrow$ $H^+ + OH^-$
Equilibrium	HCl	$\leftrightarrow$ $H^+ + Cl^-$
Salt	$FeCl_3W_6$	$\leftrightarrow$ $FeIII+3 + 3\,Cl^- + 6\,H_2O$
Salt	$FeCl_3\text{-}S$	$\leftrightarrow$ $FeIII+3 + 3\,Cl^-$
Salt	$FeOH_3\text{-}S$	$\leftrightarrow$ $FeIII+3 + 3\,OH^-$
Salt	$FeCl_3W_{25}$	$\leftrightarrow$ $FeIII+3 + 3\,Cl^- + 2.5\,H_2O$
Salt	$FeCl_3W_2$	$\leftrightarrow$ $FeIII+3 + 3\,Cl^- + 2\,H_2O$

Figure 63

When we do polymerization, producing something, the key thing is to get the kinetic models right. Otherwise, our simulations really don't tell anybody anything significant."

Figure 64

Equilibrium of 1 Mole FeCl$_3$ in 55.51 Moles Water at 25 °C and 1 atm

Comp	Moles	Comp	Moles
H$_2$O	55.5047186	FeCl+2	0.00131042
HCl	2.634e-09	FeIII+3	0.98332144
FeIIIOH$_3$	2.0841e-09	FeOH^{2+}	6.31e-06
FeCl$_3$	0.00022613	FeIIOH^{4-}	1.1102e-16
OH$^-$	9.0629e-13	FeOH+2	0.00526776
Fe$_2$OH$_2$+4	4.7779e-07	H$^+$	0.00528134
FeCl$_2^+$	0.00986047	Cl$^-$	2.97826425
FeCl$_4^-$	6.4948e-06		

Fe complexes with OH to form FeOH+2, which shifts the water dissociation to form H$^+$ ions. pH = 2.22

He showed a schematic diagram of molar volume vs. temperature (Fig. 67). Here x_c is the crystallinity which varies from zero to one, but is in general history-dependent so cannot be displayed as a state function. He then gave a viewgraph on special needs for polymer properties (Fig. 68), where, for the last item, non-state functions are defined to be those that are history-dependent.

He briefly discussed the issues and needs for solids (Fig. 69). "When one actually does modeling, with particle sizes, does the issue of nucleation, growth, —that would be an area, and I think chemical engineers did that maybe 30 or 40 years ago, but there's not been a lot recently."

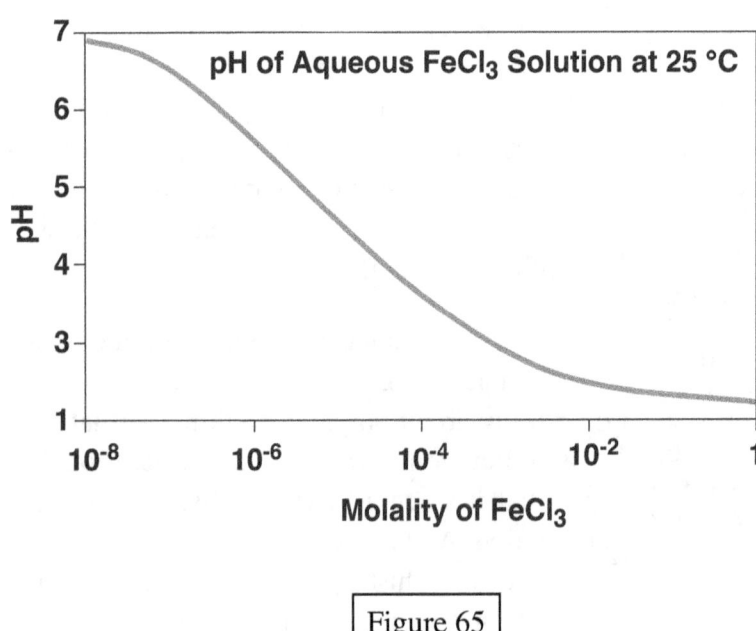

Figure 65

Mathias showed his conclusions in Fig. 70, with some further remarks on the importance of education, and on the differences between the study of conventional and unconventional properties. "I think most of the time, when we're missing something important, it's because of lack of education, and not because of lack of capability. With conventional properties, certainly there is a lot of work to be done as indicated by this conference. But, I think most of the concepts are in place. At least we know what we need to define. With non-conventional properties, with polymers and so on, I really don't know how to sit down and do a reasonable simulation, say, where the polymer is history-dependent. We just don't have the capability and the framework to describe that."

Issues and Needs for Polymers

- Polymer characterization methodologies - segment rather than component based
- Properties are history dependent
- Polymer physical property and thermodynamic models and databanks
- Polymerization kinetic models

Figure 66

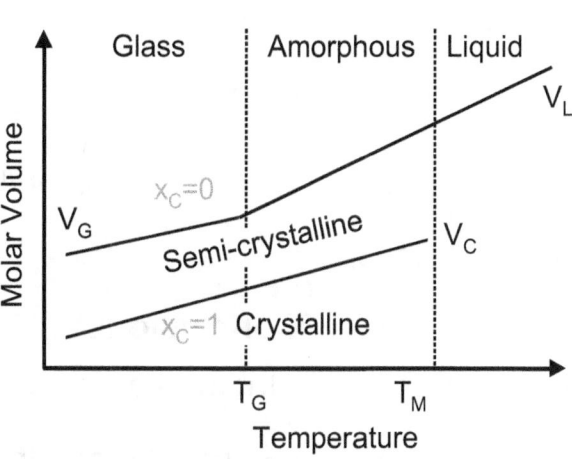

Figure 67

Special Needs for Polymer Properties

- Transport properties - viscosity, diffusion coefficient and thermal conductivity
- Phase equilibria of glassy and partially crystalline polymers
- Reaction kinetics
- Database development for non state functions

Figure 68

Figure 69

Issues and Needs for Solids

- Data, models and computational algorithms for precipitation of solids

- Prediction of morphology, habit and size distribution

- Modeling of solids processing (screening, milling, crystallizing, flotation, granulation, etc.)

Figure 70

Conclusions

- Integration of tools

- Education and training

- "Conventional" properties - concepts in place, but much work needs to be done

- "Nonconventional" properties - challenges remain even in concepts

10. Unconventional Materials

William Wakeham

The last panelist to speak was Professor William A. Wakeham, Pro Rector (Research) of the Imperial College of Science, Technology, and Medicine, London (UK). Wakeham had given a plenary talk for the 14th Symposium on Thermophysical Properties about the history and future prospects of the thermophysics profession on the previous Monday. That talk had generated considerable discussion throughout the week. He began by expressing the hope that he would not repeat that talk too much, but also that he had a point of view somewhat different from the previous panelists, in view of his career perspective (Fig. 71). "I am, of course, an experimentalist, and that means I need to defend that side of the house a little bit. I'm actually also a provider of experimental thermophysical property data, although I also play a little with theory and the applications of theory, which, in practice, means approximating it. And, I've even gotten involved in estimation procedures."

General

- Provider of thermophysical property data rather than a user
- Experimental Design
- Measurement
- Theory
- Application of theory (Approximation)
- Estimation procedures

Figure 71

He presented a classification of types of fluids of considerable interest, including "soft solids" (Fig. 72). "To pick up the point from Dr. (Paul) Mathias, the oil and gas industries will continue to be extremely important in the near future. But, I see also growth in areas connected with materials such as molten metals, as people try to make more and more sophisticated shapes, with greater and greater strength, relying on local cooling to produce particular properties in particular regions, the strength in corners, and so on, and keep the weight down while the strength is up. It's a very sophisticated modeling process of a fluid mechanics kind, in which the thermophysical properties are absolutely key. If you don't know the viscosity and thermal conductivity and density of a molten metal, which we generally don't, then you are in considerable difficulty trying to do that."

Context

- Oil
- Gas
- Molten metals
- Mixtures
- Soft Solids (Pastes)
- Inhomogeneous fluids
- Reacting mixtures

Figure 72

From his list, Wakeham also emphasized soft solids as a topic that was unlikely ever to decline in importance. "Soft solids are things like soap, food, powders, pastes of all kinds. These are everywhere. The two things you can be certain will be important in all centuries in the future are people's health and eating. The two industries that will survive everything will be food and the health industry. And, actually, they will grow. People will always want to live longer, and eat while they do it." He noted that his institution,

Imperial College of London, has chosen to concentrate on the areas of health and food, which will include several opportunities for innovative thermophysical properties research. Mixtures (including reacting ones) and inhomogeneous fluids were mentioned in passing.

Applications

- Reservoir engineering
- Refining
- Chemicals (bulk) and Pharmaceuticals
- Human tissue
- Casting processes
- Crystal Growth
- Food
- Environment

Figure 73

Wakeham listed some important future applications of thermophysical properties (Fig. 73), starting with reservoir engineering. "I talked also on Monday a little bit about reservoir engineering, and the fact that thermal diffusion, although an abstruse transport property, will turn out to be extremely important in oil reservoirs, the reason being very large temperature differences exist there. And the normally small thermal diffusion effect is enhanced by convection, and this will produce very large changes in composition in oil reservoirs. In such a volume market as the oil business, a fraction of a percent makes a lot of difference."

Wakeham proceeded to discuss in more detail those applications that he thought were most important, starting with reacting fluids (Fig. 74). He mentioned a recent conversation with a man from Sandia Labs in California about the transport properties of combusting systems, and that he was unable to tell the man how to calculate them, and that we cannot do so at present, but it is an important problem. "There are a lot of short-lived species in these things, and we don't know anything about how to treat them. In chemical engineering terms, reaction with separation of some kind is also going to be important. Reactive distillation is already with us. I think reaction in supercritical systems, when the separation may be achieved simultaneously, is going to be important."

Reacting fluids

- Combustion systems
  - Turbulent flames
  - Short-lived species
- Reaction with separation
  - supercritical systems

Figure 74

Figure 75

Inhomogeneous Fluids

- Colloids (e.g. clay suspensions)
- Aphrons
- Liquid Crystals
- Slurries (Concrete)
- 'Mushy Region' in molten metals

The next application discussed was "a more oddball thing," tissue engineering, and he reviewed the current technology for growing "pieces of people" that could be used for transplantation without large amounts of anti-rejection drugs. "Bone is rather easy to do. Lungs will be within five years. I'm sure it will be possible to grow small pieces of lung, and liver perhaps shortly afterward. Those will be grown inside a chemical reactor. It will be no different, really, than a normal chemical reactor. Plus, all the properties you need will still be there, the same sorts of thermophysical properties. At the moment in Britain, at least, doing those developments depends on an act of Parliament to change the law about whether you can use embryonic cells, but I'm certain the law will be changed, and then all these things will be possible. So, I think that's an interesting development, which this community has nothing to do with at the moment, largely."

Is experiment necessary?
- It does provide physical reality but,
  - Expensive
  - Takes time
  - Dangerous
- What might happen if we omitted the experiment?
  - Examples
    - VLE in heavy ends distillation

Figure 76

Another application he identified as important was biotechnology and membrane transport for effluent treatment and separations, where fluid/solid interaction is important. "Going back to one of my themes on Monday, connecting the process with the property will turn out to be important. The understanding of membrane transport is near zero. Membranes may be used in effluent treatment, separation at end of pipe, or in other applications such as the oxygen separation we heard about. Those things will be very important, and there's something to do in there."

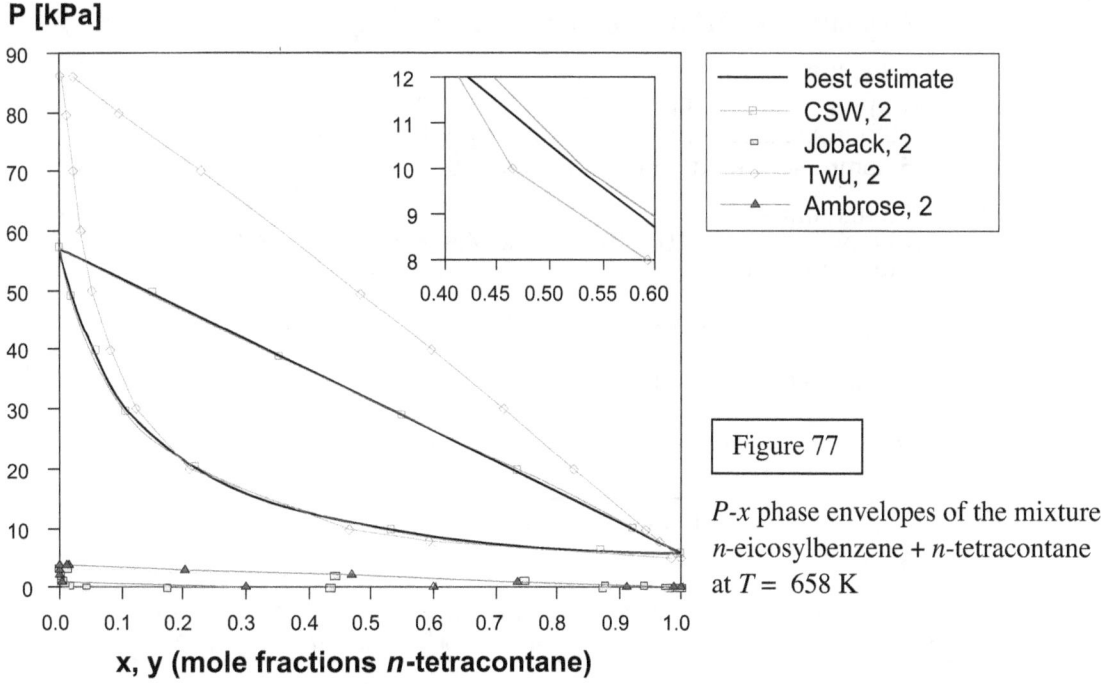

Figure 77

P-x phase envelopes of the mixture n-eicosylbenzene + n-tetracontane at T = 658 K

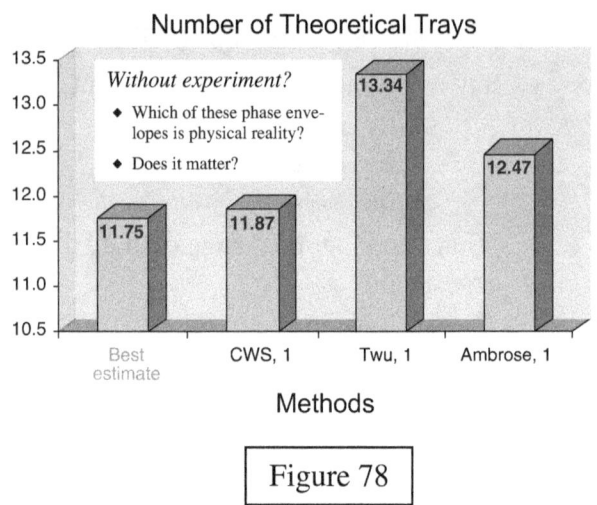

Figure 78

Wakeham went on to list various types of inhomogeneous fluids (Fig. 75), with particular emphasis on the last item, the "mushy region," which has been generally incomprehensible to the fluids community. "It is actually the region in which you have particles of solid metal combined with an as-yet-to-be frozen piece of metal. In the solids community, this is almost thought of as a thermodynamic state, which, of course, it isn't. But they measure in this region thermophysical properties. Of course, what they do discover is that it depends on how quickly you heat it or cool it or whatever, because it's not a thermodynamic state. That's a very important region." He mentioned another important aspect of inhomogeneous fluid work, multiphase flow. Even in a simple gas-liquid system, there are many empirical correlations, but no real understanding. This problem connects the fields of fluid properties and fluid mechanics.

Wakeham listed the various types of "soft solids," including pastes (food, adhesives, and cosmetics) and powders (food and cosmetics), where there are also opportunities for the thermophysical properties community. "Whatever else you may think of cosmetics—it's a very big business in the world and will grow—powders, the same. Those are areas where this community, I think, is not very active at all, and there's something to be done."

Can we possibly measure everything?

Assume:
10 properties, 10 temperatures, 10 pressures,
15 pure fluids & all their multicomponent mixtures,
5 compositions, liquid and gas phases

Total number of required measurements $= 10 \times 10 \times 10 \times 32766 \times 5 \times 2 = 3.3 \times 10^8$

Assume: 3 measurements / day

$\Rightarrow$ Man Years required ≈ 300.000

and there are many more than 15 pure chemicals in industry...

Figure 79

In conclusion, Wakeham addressed a central question of Forum 2000: Is experiment necessary? (Fig. 76). "I think we've heard a couple of arguments that may suggest it's not, because it's expensive, it takes time to do the things people really want, it's dangerous." As a counter argument, he showed one example.[8] Figure 77 displays a number of possible *P-x* phase diagrams for the binary system *n*-eicosylbenzene and *n*-tetracontane at $T = 658$ K generated by several accepted methods of prediction within a standard process design computer package. No experimental data are used in these predictions. Figure 78 shows the number of theoretical trays needed to achieve a prescribed degree of separation of a particular mixture of *n*-eicosylbenzene and *n*-tetracontane for each of the prediction schemes shown in Fig. 77. There is evidently a large variation in the distillation column that would be designed, depending on which prediction is believed.

Despite the clear need for experiment, Wakeham noted that it would be impossible to measure all properties, fluids, and mixtures of interest (Fig. 79). Even if there were only 15 pure fluids in the world, an experimental program to provide a representative sample of properties of those fluids and all of their mixtures would require 300,000 man years, which would not be possible, and in fact there are many more than 15 fluids of importance. Wakeham ended his talk with the following conclusion. "So, I think this argument about the debate between simulation, theory, and experiment is a balance between them all. You can't do away with experiment, but neither can you do all the experiments. They will have to be balanced." His final viewgraph listed the topics that he had highlighted (Fig. 80).

Conclusion
- Complex Fluids
- Inhomogeneous Fluids
- Reacting Fluids
- Balance Simulation/Theory with Experiment

Figure 80

After Wakeham's talk, and before his summary, Hanley asked if there were any immediate questions or comments from the floor. Kincaid offered the following for O'Brien: "I wondered if the new culture in West Virginia-DoE also contains the possibility of much smaller scale power generation throughout the country. It is actually conceivable that in twenty years as much as 50 % of the electrical power generation in the country could come from on-site small systems. Is that part of DoE's current planning to support that development?" O'Brien replied: "Distributed power generation is a very big concern and, perhaps most specifically, in the turbine program—the program that is just wrapping up now is the advanced turbine system—and the

---

[8] Wakeham, W. A., G. S. Cholakov, and R. P. Stateva (2001). *Consequences of property errors on the design of distillation columns*. Fluid Phase Equilib., **185**(1-2), 1-12.

product of that program was a Series H Westinghouse turbine, which is a huge turbine to generate at a central site power from natural gas at 60 % or 55 % efficiency. However, the next-generation turbine program, which is the one that is ramping up, is going to focus on smaller turbines for distributed power generation and also the virtual-demonstration idea is concerned with being able to design a whole range of turbines. So the activity won't be just through one huge turbine, but to be able to design turbines of different sizes depending upon what sort of site they're going to be located at."

Kincaid then asked to what extent DoE is supporting individual power systems for households. "I'm testing right now at the university a five- to seven-kilowatt fuel-cell system with a reformer that runs off natural gas, which would supply all of the electricity for a private home and most of its heating. I'm absolutely amazed. Its cost is coming down and it seems to me this could totally change the electrical power distribution system in the country if a few breakthroughs are made."

O'Brien replied that his expertise was in systems with turbines, but that there is also a fuel-cells program, separate from Vision 21 and a central power generation system. "The central idea (of the fuel cells program) is distributed power generation. And so that's very much a concern. And it makes planning very difficult, because it's relatively simple to plan for central power generation, but, when power generation is totally distributed, it's a very complicated and a very different problem, but it's certainly part of both the fuel-cells program and the turbine program."

11. Forum Summary

Howard Hanley

Hanley then gave an impromptu summary of the Forum based on "bullets," major themes provided by the reporters (Fig. 81). "I think both a consensus and one or two unresolved problems are still with us. The gap between simulation and experiment, I still think, is somewhat with us, but I have a feeling there is somewhat of a confusion between a process and the data itself. Now when I teach thermodynamics I always make a remark, because teaching thermodynamics is very boring to graduate students, but I like to remind them that a thermophysical property is actually a response function when you do something. If you squeeze something, the pressure tells you what happens when you squeeze it."

Hanley elaborated on distinctions that were raised: "Clearly, computer modeling is absolutely mandatory to simulate a process, but I still think that somewhere along the line you are going to have to feed in something and I go back to my rather facetious remark about chicken fat, bearing in mind the Hanford example, that's exactly the kind of information you are going to feed in. Again using the Hanford example, the gap between engineering demands and property information I still think exists and that probably will only be corrected by field work. People actually trying to find out what people want."

Forum 2000 Summary

- Gap between Simulation and Experiment (Process vs. Data)
- Gap between Engineering Demands and Property Info
- Nanoscale Demands (Complex Systems)
- From 'Conventional' to Complex
  • Polymers
  • Soft condensed matter
  • Metals
- Chemistry

Figure 81

Hanley expressed some surprise that nanotechnology was not discussed more. "Nanotechnology is probably the most, I would say, not only exciting but certainly the most current thing in the literature today. You put nano in front of anything and you are immediately in the high tech age. This is really with us, nanoscale, lithography, nanocomposites, any material on the nano length. And these are usually complex systems. They are more than one material. They are mixed up. They are organic. They are inorganic. There is a real need for characterization of these systems."

Hanley elaborated on themes from the last two speakers, Mathias and Wakeham: "The conventional systems are probably the gas systems and so on and, as Mathias said, I think the concepts are pretty well understood and computer simulation can certainly help, because the concepts are well understood. You can check what you are doing. There is almost no discussion. But, again moving from conventional to complex for example up to the nanoscale I just mentioned is something that seems to me extremely obvious. I mean, going around listening to the talks. There have been all these wonderful talks on surfactants, on colloids. People really know what they're doing, but somehow they seem to be outside the realm of 'Thermophysical Properties,' which is very, very strange to me. And as Wakeham pointed out, people aren't working on binary mixtures in the conventional systems, but they are working in multicomponent systems in very, very complicated ones. Strange!

"Soft condensed matter, which includes the topics that you mentioned, Bill (Wakeham)—food, pharmaceuticals, paint, human tissues—that to me is with us forever, because that's unlimited. There's a huge scope there. Now, in my own experience one of the problems with these kinds of things, dealing with complex liquids or soft condensed matter, is because you are not talking to the right people, because it is a multidisciplinary thing. The average physical chemist does not talk maybe to the surfactant type chemist." Hanley gave an example from his own experience: "I was concerned with clay systems and having trouble with them and I called up a company that's been making clay systems, called Southern Clay Products, since about 1905 and they solved all the problems over the phone in five minutes that took me three years to try to figure out.

"Metals is another thing and I would also include alloys in this context. Again not touched. You have ample opportunity for both simulation and for property data and for complicated analytical techniques using radiation scattering and goodness knows what. Chemistry has been bypassed. There's clearly very much with us but I don't think people know quite how to handle it, but every time chemistry came up in this last two or three hours there was a sort of nervous shuffling of the feet and nothing happened."

Hanley asked if he had left out anything, and the field of electrolytes was brought up. "Electrolytes, which comes under the category of aqueous systems, is another thing that is going to be with us forever. Again, people duck trying to solve some of those problems, but it is most definitely an area of huge interest. Multiphase systems comes under, let's call it, complex systems."

Hanley concluded his summary and called for a final, half-hour discussion period, and asked audience members to direct their questions and comments to the panelists.

12. Third Discussion Period

Kincaid began the discussion period with a comment about the role of theory. "At the bottom there should be a line that says theorists. I think this was alluded to in a number of the remarks made by the speakers concerning the problem with correlations and everything, but Wakeham, I thought, said it best. Even if there are only 15 species in the world, you can't take enough data. Somewhere along the line, somebody has to come up with some good, sound theoretical ways of making decent engineering approximations in order to get plants built and designed and energy delivered. The whole mixtures theory is very unattractive, low-profile kind of area. You need some really good mixture theorists and somebody should think about a way to really encourage people to keep working in that area. The funding agencies certainly aren't doing a particularly good job of that."

O'Brien noted that the same problem applies to simulation of all possible mixtures. "If you did three simulations a day in order to generate that data, which would be a very good number, it would still take a long time to do the simulations, so whether you get the data from experiment or simulation, there still has to be a very good way to analyze it."

Holste followed up on some issues raised by Mathias. "One of the issues that Howard [Hanley] listed was the gap between the engineering applications and the data. Paul [Mathias] mentioned the disappearance of the corporate engineering groups. Those groups performed this function in the past. Twenty years ago, those were the people that read the academic papers, realized how this applied to what needed to be done in companies, and did it. Those groups disappeared over the last 20 years in the move toward efficiency. The only thing that has come close to replacing them are the simulation companies. And that now introduces another group. And so it is a little bit different sort of relationship and it's a fairly small number of groups compared to the number of corporate engineering groups that used to be around. So a very important link of communication, which also provided input into the companies as to the reason why research should be supported, has disappeared. Nothing has replaced it and I haven't seen anybody really discussing specifically how we might approach that problem again."

"The second part of what Paul talked about is this need for education. Along with that disappearance of the corporate engineering groups what is now happening is that more and more technical decisions are made by what he referred to as a generalist, who are bachelor's-level chemical engineering graduates who probably at most have had 35 hours of lectures of thermodynamics to help them to decide whether or not the answers are correct. So what happens is when those people need help they have to find a specialist somewhere, which normally means hiring a consultant. The people I teach, maybe some of my colleagues do better, but the people that we graduate I don't think are capable of distinguishing between a competent consultant and a charlatan. As a result I think there's a lot of money changing hands to people that don't know any more than the people they're talking to and the people that are coming out can't tell the difference."

Holste said the thermophysics community must address these issues. "These are not things that can be solved overnight, but they have a lot to do with a lack of flow of money for property measurements in an organized fashion, because what happens in the second scenario is the money becomes available to solve very specific problems. So what you have is a random walk in properties with no overall coherence and no real movement forward in the science that allows us better predictions."

Holste contrasted the past and the present. "In the areas that I work, 20 years ago the industrial consortium steering committees were people that came from the corporate engineering groups. They were as knowledgeable as the academics and we would sit down together and talk about where to go. Now those groups are staffed by a different kind of people who think in terms of very short, specific projects and, with almost no communication with the people that know something about it, they simply say measure this for me. Halfway through the project you find out what they really want to do and you realize, that in thermodynamics especially, there are other ways of getting to the same answer. And after you're halfway done you finally understand what they're trying to do and you realize there's a much easier and cheaper way to get a better number for them. So, without some sort of better interaction across the entire spectrum from beginning to end, we're not going to move ahead. Paul (Mathias) is the first speaker I've seen in a long time that's even articulated any of these things." Holste received applause from the audience after his comments.

Hanley concurred. "This is why the funding is difficult to get. Because there isn't the intermediate support. Presumably, when a sponsoring agency gets a request, he or she or it can't go to a reviewer who can really give a proper review of the work. But that's a bit of the chicken and egg sort of philosophy right now."

Laesecke posed a question to the representatives of DoE: "The funding policy seems to become also more and more fragmented. The goal seems to be to fund private corporations. We've had it several times that our funding proposals from NIST were put into a return envelope and sent back to us because DoE doesn't fund national laboratories or the National Institute of Standards and Technology. We are the institution with the continuity. We are the institution which collects the data and where the expertise is. We could work against this dissipative effect of funding small, short term projects, but apparently we have no chance." He then mentioned a university that had spun off a private research organization solely in order to circumvent these funding restrictions, wondering if institutions like NIST needed to follow suit.

O'Brien said his position was such that he could not address the question specifically, but added: "I think I generally would agree with you. NIST has served a very fundamental purpose. Actually, at the higher levels I don't know what the policy is in terms of funding. I know there is an administrative problem of funding national labs versus universities versus private sectors and some decision is made in the separation of funds. For example, funding to NIST can't be competitive legally with funding to private companies. I know that in order to conform to the law that has to happen."

Cummings gave his perspective as a member of the Chemical Sciences Council, which is a group of researchers who provide input to the Chemical Sciences Program of the Office of Science. The Materials or Chemical Sciences programs of the Office of Science are where NIST proposals are likely to be submitted. The program managers in these offices at DoE are very good people who are completely stretched in every direction meeting the demands of the DoE national labs, the university researchers which they fund and the growing proportion of their portfolio which is spent on user facilities."

He elaborated on the problems of financing user facilities. "The amount of money that DoE Chemical Sciences spends on user facilities is continually growing as a fraction of the total funding within a relatively fixed budget. The DoE program managers feel that it is imperative to keep university researchers funded. They value what goes on at the universities, but they have all of these other commitments that they have to see to and it just becomes very difficult for them."

Laesecke posed a question to Poppiti to put finances in perspective: What is the amount of funds that DoE estimates is required to clean up Hanford?" Poppiti replied "Fifty-five billion (dollars) through the year 2050." Hanley replied: "But I'm going to play devil's advocate. Your point is very well taken Arno, but it's still up to the community, when the community is the people in this room, to tell the people who handle the money that they will contribute to the problem. And it's not a valid argument to say, well if you just give us one hundredth of a percent of the fifty-five billion, which is what, five hundred thousand, something like that. It's no good saying, well it's such a tiny point, it's in the roundoff, because you still have to say that you have to contribute. I just think, frankly, that the packaging is the problem."

Pellegrino said he would also like to play devil's advocate. "In actual fact people are making incredible progress without data. We have combinatorial chemistry, high-throughput screening. We are producing all the plastic we need to support six billion people in the world. You have to go where the money is. You have to go to where there is a stumbling block for the data and I thought that was part of what this particular workshop was going to try to address. Where are there stumbling blocks such as existed a hundred and fifty years ago when the first time-piece that could help you tell longitude was a world-renowned prize? Where is there a need for data that are keeping people from moving forward fast enough to satisfy the needs? Does anyone have any ideas in that direction?"

Hanley replied that such an idea was the soft-condensed-matter measurements suggested by Wakeham. Pellegrino asked, "What's the data in tissue engineering that we need?" Wakeham gave a specific example. "If you look at the way cancer tumors are sometimes removed, which is with ultrasonics or microwaves, this is all done by heating. The thermal conductivity of the tissue you heat, the heat capacity of that tissue, which is of course mostly water but not exclusively, matters and they don't know it. Sometimes it's been measured in vitro but never in vivo. That would be an interesting and quite difficult measurement."

Pellegrino asked if NIST should be doing that measurement. Wakeham said it was not his job to say what NIST should be doing, but that the problem seemed interesting. Pellegrino said that was what NIST needed. Hanley agreed, but said: "Unfortunately these absolutely black and white things come up once or twice a century and if you're lucky enough to be working in that area then you're very famous like Bill Gates. But I do agree with you, because I've worked quite a lot with companies these days and they're making money hand over fist without the need for NIST data. But I don't think that's really a valid argument in some ways." Pellegrino replied: "It is if you want to develop perspective on how to connect with future needs. What do they want to make that they can't make without the data?"

Kabelac said that things might look hopeless with the huge amount of work to be done, but added: "I would suggest something like the annex workshops would have been very useful, for example Annex Eighteen back in the past to get this problem on refrigerants done. Maybe we need some kind of structure to get these all individuals who ask for work, who are looking for smart work, to get a kind of organization, and get kind of a structure in this work. There is a lot of work to do, but there is a lot of capacity around the world, so we need to put together these things. Might that be a good idea to get some kind of annex follow up on these subjects which were listed today?" Hanley replied that people are trying, including the present Forum and efforts by AIChE (American Institute of Chemical Engineers) and in Europe, but that Kabelac's suggestion was well taken.

Thomson asked if anything concrete would come out of the present discussion. Hanley said a document would be produced, and a consensus had been somewhat forming, and added: "Well, I thought you meant sort of morally. People can take what they've heard to heart or not. At least the grievances are out in the open a bit."

Cunningham offered a suggestion about Wakeham's proposed measurements of soft tissue. "Does NIST ever try to communicate with people like NIH (National Institutes of Health) and do joint property measurements that would assist them in their budget? I think their budget's probably somewhat larger than yours and you could probably benefit from an association with that group." Hanley replied: "Yes, NIST does try, but believe it or not it's more difficult than it appears. It's also rather difficult in the times of tight money to actually start something new, but I think it's certainly something that should be explored much more." He added, however, that "turf battles" caused significant impediments.

O'Brien commented on federal interagency projects. "I spent most of my time in the Department of Energy in 'in house research' and not on the programmatic side, but in the past few years I've gotten involved in the programmatic issues and one of the things that I see happening is that all of these issues are kind of worked out informally between the agencies. The Office of Science funds this sort of research, National Science Foundation does this, NIH does this and they stay out of each other's territories by a kind of informal agreements between the program officers and they try to complement each other and a lot of it gets into turf wars. So maybe there has to be some better way to do that. To recognize that there needs to be more interagency interaction and to do it in a more formal and more powerful and binding way than just informal agreements. And maybe that's the wrong thing to do. Maybe the best way to do it is to leave it as informal arrangements between program officers."

Cummings followed up on projects between agencies. "These interagency working groups like the interagency working group on biotechnology, the one on nanotechnology, the one on information technology, all have succeeded in creating large new initiatives that all of the agencies have benefited from. It's worth keeping an eye on these large initiatives, because they do provide the funding opportunities and give you a sense of where the money is."

Jacobsen suggested that attendees perhaps should be more involved in professional "political" activities. "I think that scientists and engineers sometimes are a little averse to getting involved in the political activities and some of the things that cause these programs to occur. I think we need to recognize that we need to become involved in a way that allows us to influence some of the political decisions that are made. The time constant on some of the things that we're talking about is such that we may think that we have only two years to develop a cure or a solution to a problem, but if we look at some of the problems like the cleanup problems that have been talked about (and by the way I empathize with that situation) I recognize that there probably isn't enough money in the world to clean up all of the sites using the current technologies. So I think there is a very strong incentive, as some of you have mentioned, for providing new science, new engineering, new demonstration technologies to accomplish these activities."

Jacobsen added: "And yet the decisions that are made are often made without the involvement and without the consultation, at least proper consultation, with the science and engineering community. And that means that some of us have to get our hands dirty in some of these activities. I would be interested to know if those of you who are here from outside the thermophysical properties community feel like we've done enough or if we need to do more."

Storvick said that the problems under discussion reminded him of what has happened at universities. "When I started I thought university administrations were very helpful. They allowed me to generate ideas. They supported me. They provided me with a place to work. In recent years I've found that they have become more managers than administrators and, as you probably can understand, the university works best with an administrator that allows resources to flow from his office to the laboratory or the classroom. The good ideas always come from the assistant professor, the associate professor and has nothing to do with the (administration) office. And as soon as they become managers we lose this initiative from the bottom. We have come to fund programs not ideas. And as I sit and listen here I think this is happening on a global scale, not only on the scale of a small university where each of us might work or in a small company where often many of us work."

Harvey offered a comment on the focus of the discussion. "I think we've gotten into the politics and funding issues maybe more than we wanted and skipped past something pretty basic that Howard put up which is the need for some sort of a bridge between the basic science that most of us do and the real engineering needs like they have at Hanford. James Poppiti has this horrible tank problem and the people don't know how to model it enough to give him even an order of magnitude answer. There are probably basic data that would help to apply a much better model, but yet there's no bridge to get from one to the other." Harvey asked who would provide that bridge, NIST, DoE, a small company like Anderko's OLI, or some other group?

Poppiti said he hoped to tie three things together. "When I started at Hanford in '96, we had some rough ideas of what was in the tanks. We kind of knew what the process chemistry was. We knew it was mostly sodium nitrate. It was neutralized nitric acid and some radiochemical things and stuff in there. We found a guy down at Los Alamos named Steve Agnew and it was very simple as far as funding goes. We gave him three million dollars over about three years and he had a model and I guess it's in the area that you guys work. He modeled precipitation. Things were going in and out. Things were precipitating. Temperatures changed. Things were soluble. Some things were not and so forth an so on. Anyway, he came up with an estimate of what was in each tank.

"That was the best three million bucks I ever spent. Was he absolutely right? No, but it gave us a place to start. So then we said, based on his predictions, where are we going to learn the most from taking the next sample? Each sample we take, this is soup to nuts including the analysis, it's about a million dollars a pop. So now I had a basis to say, I've only got a certain amount of money to spend in collecting data, how do I get the most out of that next sample that I'm going to take? Where will I go?

"I think there's a real need for this on the front end to help operational people to go out and to say, where do you go to get the next thing and what kinds of things should we measure. Because it wasn't clear. We can measure radiochemicals until the cows come home, but that may not be the thing that we need to measure. I'm looking for somebody that can help direct the program into the next phase. Funding's not a problem. We can find a way to get money to whoever we need to get it to. It's just finding who the right people are that have a solution or that at least can help us get to a solution."

Hanley reiterated that a problem has to be posed and the modeler has to respond to the problem, and sometimes the problem isn't posed. At this juncture, since the allotted time had elapsed, Hanley adjourned the Forum, and thanked the panelists and the audience.

13. Conclusions

Overall, the Forum raised many questions and reached perhaps rather few definitive conclusions, but there was some consensus. While there were differences of opinion on the current capacities of simulation, it was clear that it has not at this date rendered experiment obsolete, and that experiments will need to be continued for, among other reasons, the validation of simulation predictions. Still, simulation is now an important player in the determination of thermophysical properties, and will become more so with improvements in computer speeds and algorithms. Clearly there are opportunities in new materials and properties, in microtechnology, soft solids, and elsewhere. The profession can hopefully contribute to real-life problems such as nuclear waste cleanup, but important liaisons between properties experts and practical problems, such as product-design knowledge and corporate engineering groups, have been disappearing.

A few months after the Forum was held, we solicited essays on issues raised at the Forum from panelists, audience participants, and a number of authorities in thermodynamics who were unable to attend the Forum. Thirteen essays were received, and are presented in the following section. The topics chosen and opinions expressed are those of the respective authors, who represent a diversity of experience in thermophysical properties. A recurring theme is the value and importance of experimental measurements of thermophysical properties. The authors of this Special Publication would like to thank all panelists, audience participants, and essay authors who contributed to our Forum.

14. Contributed Essays

Thermophysical Property Needs for New Technologies: Electrolyte Systems

Andrzej Anderko
OLI Systems, Inc.
Morris Plains, NJ 07950, USA
`aanderko@olisystems.com`

The importance of thermophysical properties for the design and operation of industrial processes is widely recognized. For example, the role of electrolyte properties is well known for traditional industrial applications such as separation processes (e.g., solution crystallization, extractive distillation or seawater desalination) and environmental applications (e.g., gas treatment, wastewater treatment or chemical waste disposal). However, there are several emerging areas of technology in which thermophysical properties are of central importance. The objective of this essay is to discuss the role of thermophysical properties of electrolyte systems in several selected areas of technology. In particular, we will discuss applications to corrosion simulation, natural scale formation, supercritical-water oxidation, synthesis of inorganic materials in hydrothermal media, behavior of complex organic molecules, and modeling the behavior of contaminants in soils.

Simulation of corrosion in industrial systems is an emerging new area of technology that strongly relies on thermophysical properties. Although corrosion simulation is just beginning to be practically useful, there is a strong motivation for developing accurate simulation tools in view of the high costs of corrosion (e.g., the cost of corrosion is estimated to be equal to 6 % of the annual gross revenues of the chemical process industries). In particular, corrosion simulation requires the knowledge of speciation in the system of interest. This includes pH, activities of aggressive and inhibitive ions, precipitation of sparingly soluble solids that may form films on metal/oxide interfaces, etc. The computation of speciation requires the knowledge of both standard-state properties and activity coefficients of various species. These properties are usually derived from thermodynamic measurements of vapor-liquid and solid-liquid equilibria, equilibrium potentials of electrochemical cells, etc. Although the investigation of these properties has been a classical subject of research for several decades, corrosion simulation requires data for many systems that have not been investigated in detail. In addition to speciation calculations, the knowledge of transport properties is required with particular emphasis on the diffusivity of electrochemically active species. The thermodynamic and transport properties are then used as input to electrochemical models, which represent the reactions at the metal-solution interface and transport of species to and from the interface. The need for thermodynamic and transport properties for corrosion simulation is particularly important for mixed-solvent and nonaqueous systems, for which data are much less abundant in the literature than for aqueous solutions.

Natural scale formation, especially in oil and gas production, is an area in which thermophysical properties are particularly important. Here, it is essential to accurately predict the formation of various solids that may precipitate from produced waters. The scales that are formed are very common solids such as carbonates and sulfates and one could suppose that almost everything should be known about them. However, the available experimental data are amazingly scattered and it is perilous to use them to make predictions for natural systems. The situation is particularly unsatisfactory for high-calcium brines, which are prevalent primarily in the Middle East. Therefore, there is a need for a systematic experimental program that would elucidate the properties of systems containing natural brines and solid precipitates. This would be essential for establishing meaningful computational models.

Supercritical-oxidation technology is an area in which the need for better thermophysical properties is particularly acute. The design of supercritical-water oxidation processes requires the knowledge of phase equilibria (in particular, solid-liquid and solid-gas equilibria) as well as kinetic characteristics of oxidation processes. Phase equilibria define the operational envelope within which the oxidation process is feasible. Since various oxidation processes, coupled with the neutralization of reaction products, lead to the formation of inorganic salts, the behavior of such salts in high-temperature and supercritical water is of particular importance. Recently, models have been developed for the computation of phase equilibria and thermodynamic properties of salt-water-nonelectrolyte systems. However, there is a dearth of data for calibrating such models. In particular, the behavior of multicomponent systems at high temperatures is poorly known. Also, very little is known about the transport properties of high-temperature systems that contain electrolytes. Thus, there is a need for further experimental investigation of the thermophysical properties of high-temperature electrolyte solutions. Additionally, there is a more immediate need to collect and critically evaluate the experimental information that is scattered in the literature.

An important area in which the knowledge of thermophysical properties can make a difference is the synthesis of advanced inorganic materials. For example, a variety of ceramic materials for electronic applications can be synthesized in hydrothermal media. Such syntheses offer the possibility of controlling the purity and morphology of the obtained materials while ensuring that the synthesis is environmentally benign. Thermodynamic modeling offers the possibility of optimizing the synthesis by helping to select the appropriate starting materials, their concentrations and synthesis conditions. Such models have been developed for the syntheses of several piezoelectric materials and good agreement with laboratory data has been obtained. However, the usefulness of thermodynamic modeling is limited by the availability of thermophysical data. Here, the most important data are the thermochemical properties of various aqueous species and complex solids. It is particularly challenging to estimate the properties of various solid multicomponent oxides, which may frequently contain less-common elements (e.g., zirconates, titanates, niobates, etc.). Therefore, there is need for new experimental measurements as well as for the development of techniques for predicting thermochemical properties.

Another area in which there is a need for improvement is the behavior of complex organic molecules in bulk aqueous solutions and at solid-aqueous interfaces. Here, the emphasis should be placed on developing predictive models based on molecular structure in view of the fact that the number of organic molecules is virtually unlimited. A lot of progress has been achieved in this area using both semi-empirical structure-property relationships and molecular modeling. However, there is still a need to improve the accuracy of the available methods and to increase their availability in commercial programs. For example, there is an increasing need to model adsorption of organics on model metal or metal-oxide lattices. Such modeling would be useful for the intelligent design of corrosion inhibitors. On a more basic level, prediction of the properties of organic molecules in water is necessary for a variety of environmental applications.

Another important frontier for thermophysical properties is the behavior of aqueous electrolyte systems in the environment. The interest in such systems is motivated by the need to predict the transport, fate and bioaccumulation of inorganic and organic pollutants in soils. The behavior of pollutants is governed by several physical phenomena including adsorption (i.e., ion exchange, molecular adsorption, surface complexation), phase equilibria in bulk liquid systems (e.g., partitioning between aqueous and nonaqueous organic phases) and speciation effects in the aqueous phase (e.g., complexation of metal ions, precipitation of solids, etc.) These phenomena are intimately coupled in natural environments. Recently, accurate computational models have been developed for studying these phenomena. However, further progress is limited by the availability of the necessary input data. In particular, thermodynamic data for complex organic solutes at infinite dilution in water are frequently unavailable. There are very limited data on the partitioning of organic solutes between aqueous and hydrocarbon phases. There is also a dearth of adsorption data for both ions and organics on model soils.

Finally, the role of computerized databases deserves attention. Databases serve as an indispensable link between experimentalists and industrial end users. The importance of collecting experimental data from original literature sources has been recognized for a long time and extensive databases have been collected. For example, in the area of pure fluid properties, comprehensive databases have been created under the auspices of DIPPR and by the Thermodynamic Research Center (TRC) at the Texas A&M University.[9] Collections of data for mixtures of nonelectrolytes are distributed by DECHEMA and TRC. However, no computerized databases are available for electrolyte solutions. Several, rather incomplete, data collections are available only as books. It is not clear why the status of thermophysical databases is so different for nonelectrolyte and electrolyte mixtures. Thus, there is a need for the development of computerized databases for electrolyte systems. Such databases are needed for electrolyte solutions at both normal conditions and elevated or supercritical temperatures.

---

[9] Ed. note: The Thermodynamic Research Center returned to NIST (Boulder) in September 2000. See the essay of TRC Director Michael Frenkel on p. 83.

Comments on Forum 2000, Fluid Properties for New Technologies: Connecting Virtual Design with Physical Reality

John R. Cunningham
Invensys Process Systems (formerly Simulation Sciences Inc.)
Brea, California 92823, USA
jcunningham@simsci.com

My comments are concerned mainly with the topic of *Structural Changes* facing the Thermophysical Property community. I have considered myself a member of this community for 30 years, first as an experimentalist, following with the last two decades spent as an employee of a process simulation provider.

Early on in my experience as an "industrial" thermodynamicist, the importance of sufficiently accurate physical properties seemed to be widely recognized, both by the simulation companies and our end users. As additional capabilities were incorporated into the simulation tools, and links to many of the physical property database providers were offered the availability of physical properties seemed to be taken more and more for granted. While some users have taken advantage of these new relationships, the prevailing impression or attitude of many of our customers today is that use of a simulation tool guarantees his access to sufficiently accurate physical properties. There is generally felt to be no responsibility or need for any review of basic property data or methods by the user, the assumption being that work on the tool itself has done the work necessary for any specific project. Only when a detailed analysis of a process becomes available do the problems with physical property data become visible. It has been my experience that these types of problems can crop up even in the areas where extensive property and phase equilibria data exist and well accepted process models are used.

Compounding the problem of the end user of property data sometimes abdicating his responsibility is the ongoing change in the simulation industry itself. Virtually all of the simulation industry has moved on to a wider focus of process control and industry-wide optimization. One further result of this move has been the increasing importance of derivative properties necessary for the solution of large equation-based simulations. While no more difficult than developing the property itself, it is an added level of complexity. Actual simulation is now responsible for a reduced portion of the revenue of simulation providers and is no longer an overriding concern of management. This has in turn diluted the thought and attention that is given to the underlying thermodynamic properties. The simulation companies now see themselves more as conduits, providing access to the thermodynamic community for the process engineer.

About the only suggestion I have is to reiterate in the design courses that members of our community have an opportunity to teach the ongoing necessity to validate the physical properties and thermodynamic methods used for process design.

Dynamic Compilation: A Key Concept for Future Thermophysical Data Evaluation

Michael Frenkel
Thermodynamics Research Center
National Institute of Standards and Technology
Boulder, Colorado 80305-3328, USA
Michael.Frenkel@Boulder.NIST.Gov

I have been involved in thermophysical and thermochemical data measurements, statistical and group contribution calculations as well as their critical evaluation for more than 25 years, first, working with Prof. Kabo in Minsk (Belarus), and then from 1991 at the Thermodynamics Research Center (TRC), Texas A&M University. Recently, I have been a part of the transfer of the TRC back home to NIST (TRC originated from NBS in 1942) becoming its fifth Director.

In this letter, I try to summarize the status of the critical evaluation of thermophysical property data and its possible future development.

Critical data evaluation is a very important part of chemical process design. It is also becoming accepted by the engineering community that the quality of data (essentially, defined uncertainties) is a dominant factor in determining a quality of the final product of the process simulation. On the other hand, the amount of published raw experimental data increases very rapidly, approximately doubling every 10 years. Our estimates show that currently about 700 thermophysical property data points are published daily. Critical data evaluation has always been a very time- and resource-consuming process. Taking into account new challenges related to both the quality requirements and the amount of data to be reviewed, it is difficult to imagine that presently used data evaluation procedures (we might call them static) would provide the results that industry anticipates to carry out the process design for new chemical and pharmaceutical technologies. It is an even more frustrating factor, that it is necessary to repeat the evaluation process for the same compound/property group every two or three years to update the recommended data.

As an alternative, the new concept of dynamic compilation has been offered.[10] This concept requires the development of large electronic databases capable of storing essentially all the raw experimental data known to date with detailed descriptions of the relevant metadata and the uncertainties. The combination of the databases of such a type with artificial intelligence software products designed primarily to generate sets of recommended data based on available raw experimental data and their uncertainties leads to a possibility to produce data compilations automatically to order. It is obvious that the implementation of this concept might have a very significant economic impact, potentially changing the nature of chemical process design. It is also obvious, however, that the development of the databases necessary to implement dynamic

---

[10] **Wilhoit, R. C. and K. N. Marsh** (1999). *Future Directions for Data Compilations*. Int. J. Thermophys., **20**(1), 247-256.

compilation mechanisms[11] as software tools to normalize, sort, fit the data and to generate output files directly suitable for direct processing by most popular simulation engines, is a very complicated and multifaceted engineering and computational task.

Potentially, the development of the concept might be significantly expedited by establishing archives of raw experimental thermochemical data being directly updated by those who obtain (measure) data at the time these data are submitted for publication. Certainly, the publishers and editors of the journals in the field would be very instrumental in making this happen. It is very encouraging that it seems like this particular idea is gaining some momentum now.[12] The implementation of the dynamic compilation concept would also require the development of strong international cooperation in data exchange. That, in turn, provides justification and the necessity to develop file standards for thermophysical data.

It still remains to be seen in the coming five to ten years whether the promising concept will become an everyday engineering tool. However, it seems as though all the components are falling in place for now. TRC is committed to play an active role in these developments.

---

[11] **Frenkel, M., Q. Dong, R. C. Wilhoit, K. R. Hall (2001).** *TRC SOURCE Database: A Unique Tool for Automatic Production of Data Compilations*. Int. J. Thermophys., **22**(1), 215-226.

[12] **Marsh, K. N. (2001).** *Editorial*. J. Chem. Eng. Data, **46**(1), 1.

An Academic Perspective on Experimental Thermophysical Properties Measurements

James C. Holste
Department of Chemical Engineering
Texas A&M University
College Station, TX 77843-3122, USA
j-holste@tamu.edu

Experimental studies of thermophysical properties in U.S. academic institutions have declined significantly over the last few decades because of several factors. Some of the most important of these are:

1. the perception that such thermophysical properties measurements are not "cutting edge research" discourages graduate students from entering the research field;

2. the perception that computational tools are capable of predicting properties with sufficient accuracy for process design leads to the assumption that accurate measurements are no longer necessary;

3. funding from government (both federal and state) to support long-term, systematic studies leading to fundamental advances is virtually nonexistent;

4. the movement of industry toward "just-in-time" measurements of just sufficient accuracy and range when experimental values are required for process design does not provide extensive accurate data sets.

Each of these factors has long-term implications that may not be manifested for several years, but which eventually must be addressed.

Both graduate and undergraduate students are reluctant to enter the thermophysical properties sphere because it is perceived to be unexciting in spite of its importance to successful process design. The subtle interrelationships that exist among the various thermodynamic and transport properties are not quickly grasped by most students, so, without additional research or project experiences going beyond the standard curriculum, new graduates experience difficulties in evaluating results provided by the sophisticated process simulators currently used for process design. Furthermore, the movement of graduates from the university campus into industry positions is a major (and underappreciated) mode of technology transfer. The experience acquired in analyzing experimental data allows the students to become much more competent in thermophysical properties relationships, and the experimental measurements provide excellent background for process development and troubleshooting, and for pilot-plant operations, in addition to laboratory research. Therefore, investments in thermophysical properties research should be considered as an investment in human resource development.

Computer simulations and theory both have advanced significantly in recent decades, but neither is yet capable of sufficiently accurate predictions for all substances and all mixtures.

Therefore, all three elements of the experiment, theory and simulation triad are still required for reliable process design. Theory and simulation now provide reliable answers for many simple substances at common process conditions, but shortcomings emerge for complex substances and at extreme conditions of pressure and temperature. The pertinent experimental measurements become more challenging, requiring considerable expertise and experience to provide results of sufficient accuracy to be useful. Another challenge to be met is to achieve a proper balance among the triad. Currently, simulation is emphasized nearly to the exclusion of experiment and theory. Perhaps such an imbalance is to be expected as a new field emerges, but both the theoretical and experimental challenges also are increasing in difficulty.

The lack of long-term funding (even at a modest level) for measurements threatens the ability to evolve new technology capable of the most difficult measurements. Experimental measurements in complex systems require the continued development of new techniques and instruments, as do extreme conditions of temperature and pressure. Wide range data sets are required to develop and verify predictive theories, which are much more valuable than interpolative correlations. Development of new techniques is expensive and time consuming, and failure to fund continued studies after a measurement technology is developed is an inefficient use of funding. Continuing measurements, although not inexpensive, are much more efficient than occasional measurements where continuity of expertise and experience is lost between sequences of measurements.

Demands for efficiency and increased profitability in industry have led to reduced levels of staffing and extreme emphasis on cost-reduction. One result of this trend has been the elimination of many corporate research groups and in-house specialty expertise. Historically, these groups provided the conduit for the flow of information from the academic laboratories to industrial practice. More recently, many industries have come to rely upon certain process simulators as a primary source of physical properties. As a result, when chemical systems are encountered for which the process simulators are not reliable, external sources of experimental values are required. However, the purchased measurements usually are limited to the minimum accuracy and number of values required for the immediate design task. However, such an approach does not provide for the accumulation of sufficient experimental data to support more fundamental theoretical developments that will lead to accurate and effective prediction of properties without additional measurements. Furthermore, because academic institutions are non-profit and do not accumulate cash reserves, long-term development of equipment and techniques, and the supporting research infrastructure, cannot be funded from internal sources. Also, the industrial collaborator now expects a "turn-key" system, such as computer software that provides properties at specified thermodynamic states. Such turn-key packages require more labor, therefore they are more expensive than simple measurements and require a longer time to complete.

The implications for academia and industry of these trends suggest that new modes of university/industry collaboration and communication must be developed. Generally, much closer and more frequent interaction will be required between academic and industrial collaborators. Thorny issues such as academic freedom *vs.* proprietary trade advantage, scholarship *vs.* product development, and ownership of intellectual property rights must be dealt with in new ways that are beneficial to all participants. Students will benefit significantly from closer working relations with industrial partners because the differences in working environments will be reduced, leading to better preparation of graduates for research and development careers in industry. Also, government agencies should consider the funding of thermophysical properties research at academic institutions to be an investment which develops human resources as well as generating new knowledge. Finally, because experimental measurements are critical to the continued development of predictive techniques, physical-properties data should be in the public domain, not maintained as proprietary information. It is extremely rare that the physical properties themselves lead to a true competitive advantage. However, accurate thermophysical properties are essential in our quest to increase energy efficiency, reduce adverse environmental effects, and provide for a safer chemical process industry.

Thermophysical Properties – The Nature of the Science and Art

Richard T Jacobsen
Idaho National Engineering and Environmental Laboratory
Idaho Falls, Idaho 83415-3790, USA
jacor@inel.gov

The fundamental nature of thermophysical properties research is that it is almost never the single, major focus of a project or program. Rather, accurate thermophysical properties are needed to validate or improve a system design or analysis for a particular purpose. Properties provide the basis for optimization of energy production and utilization systems using well-known working fluids and for determining feasibility of systems using working fluids that are not well characterized by measurement.

Very often, a design depends upon new or unavailable information, and the need is acute because of a tight project time schedule. The realization by an engineer or manager of the need for immediately unavailable properties is usually followed by a converging sequence of telephone calls that ends with the National Institute of Standards and Technology (NIST) or an expert known to, and referred by, NIST. Most of the time the information requested is not available on short notice or without effort by scientists or engineers. Our focus on customer needs leads us to make our best efforts to provide information to those who need it, sometimes even if it is in a confusing or unusable form. The relationship that results from such an interchange is only a small improvement over that which results from a proposal to perform the necessary research to answer the customer's question over the next three years at an estimated cost of $100,000 or more.

The value of prediction and simulation in the design and analysis of systems and processes is well established. The key to understanding the value of predicted properties to the engineering practitioner lies in the careful specification of uncertainties by the researcher making the prediction. Just as in measurement of fluid properties, the assignment of uncertainty is the responsibility of the individual or agency making the prediction. It goes without saying that predicted properties depend upon the validity of the model used, and that predicted properties validated by measurement, including those produced by direct correlation, are more likely to be accurate. Most models are based upon sound theory with some judiciously added empiricism justified by an improvement in the accuracy of calculated values. Such models are a reasonable marriage of good science and art.

The next few years of properties research will continue a trend started a few years ago. That trend is the need for characterization of substances that are not pure substances in the sense of well-defined, clean laboratory-prepared samples. We need to know the properties of mixtures like methane hydrates in silt at the bottom of the Gulf of Mexico or in permafrost in Arctic regions, and of colloids including radioactive substances in subsurface water transported through porous or fractured media below buried waste. We need not only equilibrium properties of such mixtures but also transport properties. The available transport properties for both fluids and solids are less prevalent than equilibrium properties and generally are considerably less accurate, a fact not necessarily widely understood by users of properties in design work. The best science is helpful for this work but incomplete. Even computing resources are at times inadequate for very complex models. The available experimental data sets are not yet comprehensive enough to allow practical models to be validated.

One of the challenges given to students in a thermal systems design class is the design of a heat pump small enough to fit in a lunch box yet capable of heating and cooling a 3000 square foot (275 m^2) multi-story residence. The purpose of the assignment is to engage the students in a thoughtful comparison of current residential systems to the postulated design with the hope that they will bridge the gap between modern science and what may be possible in the next few years. The idea is not so far-fetched if they allow themselves to look backward in time, even in their own life times, to advances made in automotive technology, in digital computing, in communications and in many other fields. The students usually conclude that the design is impossible or marginally possible, and that it will take a lot of time and cost a lot of money to determine whether the idea is worth pursuing.

A related motivation building on the example above is a realization that scientists and/or engineers need to engage in the social, political and economic processes needed to bring the best science and technology to reality with sufficient funding and public support to help solve real problems such as the current, and possible future, energy crises. The role of properties of fluid systems including natural gas, methane hydrates and subsurface contaminants and the related science is very important in most modern engineered systems. That role will be much clearer when our national priorities are defined by a national energy policy and the concomitant federal investment has been identified. The roles of scientists and engineers in the political processes must be increased, if the results of national research programs are to be credible with the public from technical and scientific standpoints.

The need for continued research in properties of fluids and solids can best be met by strategic partnerships between scientists and designers and manufacturers of both fluids and system components. The strategic nature of such partnerships requires both trust and management of risk by the partners, but the economic advantages of collaboration at all levels, including international collaboration, are significant. It is probably true that no commercial advantage for a product or process has been totally dependent upon making fluid property values trade secrets, yet companies sometimes insist on protecting property measurements or predictions as such.

In conclusion, the future of properties research depends upon a focus by the researchers meeting customers' needs in a timely and cost-effective manner using the best science and art available. Beyond those perhaps obvious statements, the business case must be made for any new research that involves a significant cost, and the considerations of public health and the environment may be even more relevant than the science in the success of current and future work.

Micro, Nano, Pronto, Combi - What Thermophysicists (Can) Learn from Genomics

Arno Laesecke
Experimental Properties of Fluids Group
National Institute of Standards and Technology
Boulder, Colorado 80305-3328, USA
`Arno.Laesecke@Boulder.NIST.Gov`

The need for experimental data has been emphasized repeatedly during Forum 2000 and in the essays contributed to this publication. Perspectives for experimental methods and techniques were hardly mentioned. Is there a potential for innovation in thermophysical properties measurements? May significant new developments be expected in this area or has experimental science reached a stagnation point? This contribution is intended to fill the gap in the discussions by providing an outlook for future experimental directions.

The year 2000 marked a historic scientific achievement: the mapping of the human chromosome 21.[13] The total length of the DNA sequence reported is 33.55 million base pairs. Only a couple of decades ago it appeared questionable to analyze this volume of information on a human time scale. The concerns about the viability of measurements to acquire knowledge of thermophysical properties, as expressed during Forum 2000 by panelists Cummings (Fig. 5) and Wakeham (Fig. 79), as well as elsewhere,[14] are based on similar time-scale (and cost) arguments. The question is whether or not the breaking of this barrier in genomics provides orientation for the future development of thermophysics? Although both fields appear unrelated at first, clues can be derived indeed from the developments that made the sequencing of the human chromosome 21 possible after all.

The main carrier for the breakthrough in genomics was the miniaturization of analytical instruments[15] and the concurrent acceleration of measurements. The acceleration has two components. One consists of faster response times and reduced inertias of small systems, for which panelist TeGrotenhuis gave some examples at Forum 2000 (Fig. 18). The second component multiplies the speed gain through miniaturization by operating small systems in parallel. Such parallelization is familiar from recent developments of computer systems architecture and algorithms.[16] In fields other than information processing, parallelization has led to even more profound shifts in methodology. The combinatorial approach of synthesizing and screening hundreds or thousands of compounds or materials rapidly and automatically in order to explore large parameter spaces that control desired properties has significantly changed the way research is

---

[13] **The Chromosome 21 Mapping And Sequencing Consortium (2000).** *The DNA Sequence Of Human Chromosome 21.* Nature, **405**(18 May 2000), 311-319.

[14] **Deiters, U. K., M. Hloucha, and K. Leonhard (1999).** *Experiments? — No Thank You!* In T. M. Letcher (Ed.), Chemistry for the 21st Century, pp. 187-195 (Oxford: Blackwell Science).

[15] **Kricka, L. J. (1998).** *Revolution on a Square Centimeter.* Nature Biotech., **16**(June), 513-514.

[16] **Topping, B., J. Sziveri, A. Bahreinejad, J. Leite, and B. Cheng (1998).** *Parallel Processing, Neural Networks And Genetic Algorithms.* Adv. Eng. Software, **29**(10), 763-786.

carried out in chemistry, materials science, and in the life sciences.[17] The gravity of this methodological shift should not be underestimated. While the inertia of the instrumentation continues to decrease by sustained advances in miniaturization, reduction of the inertia reaction of the researchers to embrace such cultural changes may not always follow suit.[18]

In parts of thermophysics, the combinatorial approach is actually well established, although it has not been known by that term. The "blind numerology of curvefitting"[19] experimental data by adding polynomial density terms of higher and higher degree to the virial equation of state was systematized by a hybrid strategy of linear stepwise regression and evolutionary optimization. With this strategy, it became possible to screen banks (libraries) of several hundreds or thousands of functional terms for subsets with typically 30 to 60 members that represent a set of data with the greatest statistical significance and with the highest accuracy. For about two decades, this combinatorial technique has been the main tool to develop multiparameter equations of state.[20]

Experimental thermophysics is still dominated by traditional measurements of fluid properties on sample volumes between 100 and 3 ml over wide ranges of the process variables pressure and temperature. Such complete characterizations should continue for a number of key reference fluids, pure and binary systems, with unique features. These fluids should be carefully selected and the burden of the time-consuming measurements should be shared in international collaboration.

There are strong indications that the broad sweep of microtechnology has now reached experimental thermophysics. As a result, breakthrough innovations in thermophysical properties measurements will occur over the next five to ten years. Additional thrust will be generated by the growing importance of microfluidic systems vs. microelectromechanical systems (MEMS) for the simple reason that fluid flow is the easiest way to transport material on the microscale. Nature is the best example: aqueous solutions are the main transport media in living organisms.

The manipulation of fluids in microscale devices requires sensors of their thermophysical properties. These cannot be developed by simply scaling down conventional devices. Consider, for example, a falling-body viscometer. Below a certain diameter, the sphere may no longer move in the viscometer channel due to buoyancy and/or due to stiction to the wall. In addition, this discontinuous technique is incompatible with continuous flow, which will be the predominant mode of operation for microfluidic circuits. New types of sensors and on-chip measuring devices are a pressing need for the widespread deployment of microfluidic systems beyond

---

[17] **Jandeleit, B., D. J. Schäfer, T. S. Powers, H. W. Turner, and W. H. Weinberg (1999)**. *Combinatorial Materials Science and Catalysis.* Angew. Chem., Int. Ed. Engl., **38**, 2494-2532.

[18] **Borman, S. (2000)**. *Combinatorial Chemistry—Redefining the Scientific Method.* Chem. Eng. News, **78**, 53-65, and **Dagani, R.** *Materials À La Combi. Ibid.*, 66-68.

[19] **Leland Jr., T. W. and P. S. Chappelear (1968)**. *The Corresponding States Principle. A Review of Current Theory and Practice.* Ind. Eng. Chem., **60**(7), 15-43.

[20] **Span, R. (2000)**. *Multiparameter Equations of State: An Accurate Source of Thermodynamic Property Data.* (Berlin, Heidelberg: Springer-Verlag).

the laboratory[21, 22] On the other hand, the fundamental properties of fluids in extremely small geometries are largely unexplored. Their study will almost certainly lead to novel property sensing techniques, which will be also influenced ("shaped") by the microfabrication process. Examples are the recently presented microsensors for thermal conductivity, heat capacity, and viscosity measurements.[23] Thermophysicists should engage in an active dialog with microfabrication specialists to offer their expertise and to learn about new possibilities.

Miniaturization of thermophysical properties sensors will add a third combinatorial component to the acceleration that occurs due to faster response times of small systems and due to their parallel operation. A fluid sample can be sent through a series of such sensors, each of which measures a different property. The time needed to characterize a pure fluid or mixture would be once more massively reduced. Such multi-property apparatus were proposed around 1970 but did not find widespread application on the macroscale. On the microscale, such integrated instruments are known as Micro Total Analysis Systems (μTAS) and they have become the topic of a dedicated series of conferences.[24]

These developments will impact the metrology of thermophysical properties of fluids in the near future. Traditional process-variable oriented measurements of "one fluid at a time" will retain their reference character, but they will be complemented by automated composition-oriented microscale measurements of "many fluids at once." The "many fluids" may be synthesized in combinatorial reaction systems or they may be delivered by automatic mixture preparation units. Even if these measurements will be conducted initially on liquids at atmospheric pressure and in a near-ambient temperature range, they will allow scientists to probe molecular interactions of many more compounds and mixtures than has been possible before. Shifts in the development of models for these data will follow the shift in experimental methodology. The reductionism of assigning "effective" energy and length scaling parameters of elusive (E. Mason) model fluids to any molecular structure will be superseded by quantitative structure property relationships (QSPR). Such models[25] incorporate a high sensitivity to molecular architecture, which leverages into better predictive capabilities for properties of compounds not yet measured. There are good chances that the outlined developments will lead to a rediscovery of measurements in thermophysics, as is happening in other scientific fields.[26]

---

[21] Jensen, K. F. (2001). *Microreaction Engineering—Is Small Better?* Chem. Eng. Sci., **56**(2), 293-303.

[22] Giordano, N. and J.-T. Cheng (2001). *Microfluid Mechanics: Progress And Opportunities.* J. Phys.: Condens. Matter, **13**(15), R271-R295.

[23] van Baar, J. J., R. J. Wiegerink, T. S. J. Lammerink, G. J. M. Krijnen, and M. Elwenspoek (2001). *Micromachined Structures for Thermal Measurements of Fluid and Flow Parameters.* J. Micromech. Microeng., **11**(4), 311-318.

[24] Van den Berg, A., W. Olthuis, and P. Bergveld (eds.). *Micro Total Analysis Systems 2000.* Proc. μTAS 2000 Symposium. Enschede, The Netherlands, 14-18 May, 2000 (Dordrecht: Kluwer Academic).

[25] Suzuki, T., R.-U. Ebert, and G. Schüürmann (2001). *Application of Neural Networks to Modeling and Estimating Temperature-Dependent Liquid Viscosity of Organic Compounds.* J. Chem. Inf. Comput. Sci., **41**(3), 776-790.

[26] *18th IEEE Instrumentation and Measurement Technology Conference: Rediscovering Measurement in the Age of Informatics.* Budapest, Hungary, 21-23 May, 2001. IEEE: Piscataway, USA.

Improving the Physical Properties Infrastructure for Industry

Alvin H. Larsen
1480 Chandellay Drive
St. Louis, Missouri 63146, USA
ahlars@swbell.net

George H. Thomson
Design Institute for Physical Properties (DIPPR)
2308 Brother Luke Place
Santa Fe, New Mexico 87505, USA
ghthomson@cnsp.com

Industry needs a continuous flow of accepted physical property data for a wide variety of its operations. These include:

1. *Process engineering*. Accurate data are required for design and optimization of processes for production of products as well as for reduction and treatment of wastes. For example, the economics of a heat-intensive process can change considerably if the price of natural gas rises, as it did recently from $3/MBtu to $9/MBtu. It is then very important to use accurate heat capacities and enthalpies in the process calculations.

2. *Governmental regulations*. Regulatory agencies often do not have reliable physical property data in-house and do not know where to find them. Obtaining them from less-experienced consultants can lead to bad values. One large chemical company recently estimated that it will save millions of dollars by not having to cover wastewater treatment ponds which it would have had to cover if incorrect values of Henry's Law constants had been retained in the EPA's CHEM9 and WATER9 emission models.

3. *Custody transfer*. Many liquids transported in pipelines are still metered by volume, but bought and sold by mass. Accurate knowledge of liquid densities and their temperature dependence is required to convert volume to mass.

Ongoing needs for new data on conventional properties of fluids involved in chemical processes are magnified by rapidly changing technology, changing customer demands, globalization of chemical facilities, environmental concerns, and stringent time and cost pressures for getting new products to market and new capacity for existing products on-line. As a result, industry does not want to spend the time and money it takes to get accurate physical property data. In addition, physical property measurements are not "fashionable" and are not likely to be funded by government agencies, so they are not popular academic subjects. Process engineers are now less likely to have expertise in thermophysical properties, and less likely to have access to physical property specialists and laboratories within their own companies. The assumption is that users can buy whatever physical properties they need whenever they need them, so they are coming to rely on independent databases, simulation software companies, consultants, and commercial laboratories, which we may term the physical-properties infrastructure.

This infrastructure has been built piecemeal, over decades, and some of it is incomplete or "temporary." The number of available databases and commercial laboratories has also decreased in recent years, perhaps under the pressure of tight money and a perceived lack of need for data. There are many gaps in the available databases. Existing data are often questionable or inconsistent. Property-estimation methods are not as accurate or as widely applicable as one would like. New processes may require data for new components, for which no data are currently available. Engineers are faced with using whatever data or estimates they can find. If the infrastructure is improved, adequate data and predictions are more likely to be available when needed.

Process engineering provides a good example. The physical property infrastructure is important to process engineers because they use measurements, models and simulations, which are all abstractions of the reality of complex processes, in order to make predictions of process behavior. Direct measurements are often made of process variables for control purposes, but these measurements are never adequate for developing a complete process model, much less for designing the process in the first place. Therefore the process model must be simplified to make it tractable, while retaining the essential features. Key measurements are made of simplified variables, such as pure component properties and binary interactions, to develop the model. Then the model must be validated by comparing its predictions with process behavior. Improvements in the infrastructure will diminish the need for additional measurements and improve the quality of process models.

Advances in computing technology have made possible on-line, real-time models of complete processes for process control and optimization. These models include relevant thermophysical properties, and provide opportunities for striking reductions in cost. Development of such models requires expertise in dynamic simulation and real-time programming tools as well as physical properties. It is not feasible for process engineers, or even for physical property specialists, to keep abreast of advances in databases, correlation programs, simulation algorithms, and programming methods, let alone develop and maintain their own software tools. They must rely on the physical properties infrastructure.

The challenge is to convert advances in technology into improvements in the infrastructure, so they will be available to engineers for use in project work. Legacy proprietary software tools are increasingly difficult to maintain, so there is no realistic choice other than to embrace new technology and capabilities as they become available. Awareness of advances is important. Industry would like to take advantage of new tools and methods, as well as new data, but must know about them. Lively communication will also help minimize duplication of effort. Relationships between industry and the infrastructure and among the various parts of the infrastructure must be strengthened to make this conversion possible. The purpose of this paper is to offer some suggestions for addressing these problems.

First, the case must be made for the economic value of accurate physical-property data, to provide incentive for both providers and users. The existing commercial laboratories which supply these data must be supported in times when demand is low, however. If not, they will not be there when demand is high.

Second, improvement in the physical-properties infrastructure requires improvement in estimation methods, which are often used to supply missing data. While there has been considerable improvement in these methods in recent years, it is interesting to note that classical group contribution methods still produce better results than some newer, more sophisticated methods. The newer methods still need to be extended to more complex substances, simplified, and qualified so the user will have a good idea of what accuracy to expect. An example of an attempt to improve the state of the art of one type of prediction method is a workshop on Prediction of Thermophysical Properties of Fluids by Molecular Simulation, organized by NIST in cooperation with several chemical companies and universities.[27]

Third, existing commercial physical property databases must be maintained and enhanced. Data must continue to be carefully evaluated as they are added. It may be desirable to exchange data or to combine some of the existing databases. These data cannot be distributed at no cost since collection and evaluation take time and skill. Values of physical properties of the most common chemicals are not always the same in different databases. Accepted values for a "core" group of the most common chemicals are needed. This should be a cooperative effort among the creators and managers of the different databases, but might well be coordinated by NIST.

Fourth, standards for electronic transfer of data are needed. Process simulators require large quantities of physical property data and require that they be delivered quickly. Manual transfer of data is tedious, subject to error, and slow. Electronic "cutting and pasting" is also slow. A standard for electronic transfer of physical property data between computing programs and databases is under development. Another standard for the transfer of data from authors to journals and to databases is also under development. It is important that the people and organizations using large quantities of physical property data participate in the development and maintenance of these standards.

Finally, future significant physical property measurement or collection projects should be collaborations among several organizations. It would be useful to have a group to coordinate these efforts and also to speak for the physical property community. Th. A. Manuel,[28] immediate past president of the Council for Chemical Research, proposed the formation of a committee to coordinate work on the long-term technical needs of the industry, to go further with the Vision 2020 program.[29] The physical property community would benefit greatly from a group of this sort. Such a coordinating group would present the case for physical properties to funding agencies and regulatory bodies. It should consist of representatives from industry, academia, government laboratories, and perhaps trade organizations and professional societies.

---

[27] *Workshop on Predicting the Thermophysical Properties of Fluids by Molecular Simulation*, June 18-19, 2001, NIST, Gaithersburg, Maryland, USA. Detailed information about the workshop including presentation files is available at http://www.ctcms.nist.gov/~fstarr/ptpfms/home.html.

[28] Dr. Th. A. Manuel is also the chairman of the NIST Visiting Committee on Advanced Technology. His term expires on January 31, 2002. For further information see http://www.nist.gov/director/vcat/manuel.htm.

[29] *Uncertain Road for Vision 2020. Chemical Industry Partnership Seeks Rejuvenation*. Chem. Eng. News, Feb. 26, 2001, p. 10.

The physical properties infrastructure has an important role in connecting design with reality. Progress in achieving the objectives indicated here will improve the infrastructure and contribute to making the chemical industry more efficient and competitive.

Comments on the Status of Physical Properties for Chemical Manufacturing

Paul Mathias
Aspen Technology, Inc.
Ten Canal Park
Cambridge, Massachusetts 02141-2201, USA
Paul.Mathias@aspentech.com

I am honored to have the opportunity to add my personal comments to this publication. My comments discuss the status and needs of physical properties from the perspective of chemical manufacturing. My comments are a personal perspective, gained from over 20 years of applied experience.

What do I mean by " the perspective of chemical manufacturing?" It covers the areas of process and plant simulation, on-line optimization, control calculations and research on new-product development. An important characteristic of these activities is that they are performed by sophisticated technologists, typically process engineers and research engineers, but professionals who more and more often have no specialization in thermophysical properties.

Computers are essential to chemical manufacturing. These are powerful, but everyday, computers. The computer on which I do my process modeling is the same laptop on which I write my reports, the one I work on while traveling and the one I use for demos at customer sites. It is a computer similar to the one you and I buy for our homes, so that our children can do their homework and surf the net. Supercomputers have a strong role in research, but a lesser role in chemical manufacturing. Computers used in chemical manufacturing are powerful, ordinary and ubiquitous.

Most of the successes in physical properties have been with relatively simple systems, which I refer to as "conventional systems." By conventional systems, I mean mixtures that do not contain electrolytes or polymers and that exhibit phase behavior limited to vapor-liquid-liquid-solid phases (as opposed "complex" behavior, such as self assembly or glass transitions or history-dependent properties). The properties of conventional systems are covered in books such as "The Properties of Gases and Liquids," by Poling, Prausnitz and O'Connell.[30]

The successful correlative models for conventional systems have come from insightful and clever, but simple, approximations. Variants of the van der Waals equation of state describe high-pressure systems; local-composition approximations by Wilson (and its derivatives, such as NRTL and UNIQUAC) serve as the basis of most successful activity-coefficient models; and similar methods, which have stood the test of time, serve for other properties. The power of these methods derives from their bold, elegant approximations, their longevity and their adherence to the framework of the laws of thermodynamics (i.e., thermodynamic consistency). Databases of properties have been an essential link in the development and deployment of accurate and reli-

---

[30] **Poling, B. E., J. M. Prausnitz, and J. P. O'Connell (2001).** *The Properties of Gases and Liquids*, 5th ed. (New York: McGraw-Hill).

able models; this includes the measurements themselves as well as the evaluated compilations of data.

The most successful estimation methods have resulted, surprisingly, from simple group additivity methods, such as the UNIFAC correlation. Theoretical methods, based upon quantum mechanics and statistical mechanics, have not yet contributed significantly to chemical manufacturing.

Widely applicable correlations, extensive data compilations and relatively simple estimation methods have enabled widespread use of simulations by generalist engineers and scientists.

What are the future challenges for the modeling of the physical properties of conventional systems? My personal opinions are highlighted below:

1. *Education*. Education must change as engineers routinely use complex software to model systems. They must be taught how to determine which models are applicable, how best to develop optimal model parameters, whether the correlations are supported by experimental data and what is the uncertainty of the estimations. The "feel" that experts gained over many years and through many, painful mistakes must be conveyed and absorbed in the modern undergraduate curriculum.

2. *Data*. Data compilations remain the greatest need. Even for conventional systems, the need far exceeds the available databases. For example, there are several million known chemicals, while the DIPPR 801 project provides data for 1,635 commonly used chemicals; of course the requirement for mixture data exponentially increases the data needs. My comment is not intended to diminish the important contribution of data-compilation efforts such as DIPPR 801, but rather to highlight the work that remains to be done. Transport properties are becoming increasingly important, but little attention has been devoted to them.

3. *Theory*. Theoretical methods can, in principle, predict the properties needed for chemical manufacturing, but the accuracy has remained elusive. Theoretical methods are those based upon *ab initio* quantum mechanics as well as many-body-statistical mechanics. The prediction of gas-phase thermochemical properties by *ab initio* quantum mechanics is a notable success. The theoretical methods should be more widely accessible, so that the general experience will grow and mature. The challenge for theorists is to provide software that will be used routinely by industry. Perhaps what is needed is a hybrid approach that will calculate perturbations in properties from a known reference, leveraging the available data compilations.

Future efforts must focus on complex systems. While the conceptual framework is more-or-less established for the modeling of conventional systems, many problems and opportunities are evident for complex systems. I close this brief commentary with a sampling of the complex problems I have encountered recently:

1. *Polymers*. How to describe the solubility of gases in semi-crystalline systems? How do

the rheological properties change with shear? How do properties change with the molecular-weight distribution, co-monomer content, and branching of the polymer?

2. *Electrolytes.* What speciation really exists and what speciation is necessary to develop an engineering model? What is the concentration limit of the electrolyte models? When I encounter an inorganic solid, what is its thermodynamic solubility and minor impurities inhibit its dissolution? How do we model a property such as electrical conductivity?

3. *Self-Assembly.* How do we model the critical micelle concentration? Can we extend the models used for nonideality in fluids to describe self-assembly in liquids?

4. *Biological Systems.* How do we best model partitioning in two-phase aqueous systems? How is the folding of a protein affected by its medium?

The above examples are a typical, but of course a partial, list of the complex problems that engineers will increasingly have to face in chemical manufacturing. The challenge of the property specialists is to develop the practical data and tools to elucidate the complex problems and to educate non-experts to deploy the solutions.

Simulations and Sensitivity Analysis

Ray Mountain
Physical and Chemical Properties Division
National Institute of Standards and Technology
Gaithersburg, Maryland 20899-8380, USA
Raymond.Mountain@NIST.Gov

I have been at NBS/NIST for 37+ years, starting as a postdoc with Mel Green in 1963. I have been actively involved with molecular simulations (Monte Carlo and molecular dynamics) as a tool to study the properties of liquids and solids since the 1970s, when it first became possible to use these methods to evaluate theories based on statistical mechanics.

In this letter, I share some thoughts on some steps that might/should be taken to encourage the use of molecular simulations to satisfy special data needs. These suggestions are presented in the hope of generating some dialog on how the simulation community could convince potential users of simulation data that simulation results have value. The potential users, who are usually not specialists in simulations, include engineers screening fluids for specific properties, engineers seeking alternatives to expensive (and perhaps hazardous) measurements, and data compilers seeking to fill gaps in data sets. Before these persons will seriously consider using simulation results, it will be necessary to build their confidence that simulations can add (and not subtract) value.

There are some things that the simulation community can do now and some that will take more time. For starters, our results should include a conservative set of uncertainty estimates based on several factors. The first factor is the sampling/statistical uncertainty.[31,32] This is an example of a Type A uncertainty.[33] The second factor is the degree of agreement with existing experimental results, if the data exist. This is an example of a Type B uncertainty. The third factor is the sensitivity of the results to the potential/force field parameters. I will expand on this third factor below.

Sensitivity analysis should be an integral part of potential function/force field development and of the description of the potential function. It is important for the user to understand how sensitive interesting properties, such as the pressure and the energy, are to the parameters of the potentials used to obtain the properties. An introduction to the approach for static properties is found in studies by Wong and coworkers.[34,35,36] There it is shown that within the framework of the canonical ensemble (constant NVT), the sensitivity coefficients, defined below, can be

---

[31] **Allen, M. P. and D. J. Tildesley (1987).** *Computer Simulation of Liquids*. (New York: Oxford University Press) pp. 191-198.

[32] **Frenkel, D. and B. Smit (1996).** *Understanding Molecular Simulation*. (San Diego: Academic Press) pp. 377-384.

[33] **Taylor, B. N. and C. E. Kuyatt (1994).** *Guidelines for Evaluating and Expressing the Uncertainty of NIST Measurement Results* (National Institute of Standards and Technology, Technical Note 1297, Gaithersburg, MD, http://physics.nist.gov/Document/tn1297.pdf).

obtained from a single simulation. The presentation in those references should be augmented by including estimates of the sampling uncertainty.

The first-order sensitivity, $\delta\langle F\rangle$, of a property $\langle F\rangle$ to the various potential parameters λ_i is

$$\delta\langle F\rangle = \sum_i \frac{\partial\langle F\rangle}{\partial \lambda i}\delta\lambda_i ,$$

where the angular brackets, $\langle ...\rangle$, indicate an ensemble average of the enclosed quantity. For the canonical ensemble with Hamiltonian H, it is easily shown that

$$\frac{\partial\langle F\rangle}{\partial\lambda_i} = \left\langle\frac{\partial F}{\partial\lambda_i}\right\rangle - \frac{1}{k_B T}\left[\left\langle F\frac{\partial H}{\partial\lambda_i}\right\rangle - \langle F\rangle\left\langle\frac{\partial H}{\partial\lambda_i}\right\rangle\right].$$

The first term describes the explicit sensitivity of the property to the parameters, and the second term contains the ensemble sensitivity of the property. These are quantities that can be generated in a single simulation. Note that the use of other ensembles, such as the microcanonical ensemble or the constant-pressure ensemble, may involve different expressions for the ensemble part of the sensitivity coefficients.

Sensitivity analysis for transport coefficients is likely to be a much more demanding task than it is for static quantities, since the variation of time correlation functions, or of the equivalent Einstein relations, will probably require several simulations as well as long simulation runs.

To reiterate, the simulation community will have to do some selling if our methods are to have a serious impact on the data community. Providing the uncertainty estimates is an important step in the process.

One possible part of the selling process is that future Thermophysical Property Symposia include simulation sessions devoted to the documentation of successes and failures of simulations to generate useful, reliable information. This would be a part of the confidence-building, educational effort needed for this enterprise.

---

[34] Zhu, S.-B. and C. F. Wong (1993). *Sensititivy Analysis of Water Thermodynamics*, J. Chem. Phys., **98**(11), 8892-8899.

[35] Zhu, S.-B. and C. F. Wong (1993). *Sensitivity Analysis of Distribution Functions of Liquid Water*, J. Chem. Phys., **99**(11), 9047-9053.

[36] Wong, C. F. et al. (1998). *Sensitivity Analysis in Biomolecular Simulation*, in Reviews in Computational Chemistry, K. B. Lipkowitz and D. B. Boyd (eds.), Vol. 12, pp. 281-326.

The Importance of Experimental Measurements

James D. Olson
The Dow Chemical Company
Research and Development Department
Technical Center, 740-3107
South Charleston, West Virginia 25303, USA
`olsonjd@ucarb.com`

"Face the facts: you cannot get something from nothing. Do not expect magic from thermodynamics. If you want reliable results, you will need some reliable experimental data." – John Prausnitz[37]

Of the three routes to thermophysical property data (retrieval from literature, estimation from correlation or theory, experimental measurement), experimental measurement is the slowest, most painstaking, and most expensive. Many have noted the decrease of experimental thermophysical property research in the last 50 years, perhaps inversely correlated with the advent of powerful and accessible computers. When are thermophysical property data measured and why?

In industry or contract laboratories working for industry, experimental measurements are undertaken to provide the key data needed for a specific design problem. These data are important proprietary intellectual property of the company measuring the data. For that reason, companies do not generally publish these data except as required in patents or safety literature such as Material Safety Data Sheets. Government laboratories measure data for specific industries whose needs are identified as within the mission of the government agency. These data represent "leveraged" research within the public domain. In university laboratories, however, data are measured as part of individual investigators' research programs governed by the broad constraints of funding agencies, which are the university or say, NSF, NIH, DoE, and DoD. University data, like government lab data, are published in peer-reviewed public-domain ("open") journals.

A key problem noted by Bird[38] and others is that "careful measurement of thermophysical property data by university researchers is no longer recognized as a significant scientific contribution by funding agencies and university evaluation committees." One suspects that outstanding experimental thermophysical property researchers such as Sage and Lacey, Kay, and Streett (whose data appear in textbooks and advanced treatises) could not get tenure in today's universities!

There is an important value of measured data (other than its immediate short-term use) that should justify to funding agencies and universities that experimental measurement research

---

[37] **Reid, R. C., J. M. Prausnitz, and B. M. Poling (1987)**. *The Properties of Gases and Liquid,* 4th ed. (New York: McGraw-Hill).

[38] **Bird, R. B. (1996)**. *Restore the Right Priorities.* Chem. Eng. Prog., **92**(October), 80-83.

is a legitimate and equal partner with the currently more popular research activities of molecular simulation and correlation. Measured data are: (1) the raw material from which rapid, easily used, and cost-effective estimation methods are derived and (2) the ultimate validation of molecular simulation prediction methods. This self-evident value was emphasized in the Forum 2000 discussion by William Wakeham who pointed out "that we can never conduct all the necessary experiments." He called for a balance between experiment, theory, and molecular simulation.

A good example of longer-term use of measured data is continuing development of the binary interaction matrix for the UNIFAC group contribution estimation method for liquid-phase activity coefficients and for equation-of-state interaction coefficients. The original UNIFAC parameter matrix[39] contained about 42 % empty cells; continuing vapor-liquid equilibria measurements have decreased the empty space. For a recent example, see Nölker and Roth.[40]

Below are some specific suggestions for thermophysical property data measurements that would certainly be significant research projects:

- Any property measurements on "mixed-moiety" (multi-functional) chemicals – An important example is monoethanolamine (2-amino-ethanol). MEA properties cannot be estimated by any of the common homologous series methodologies. Another industrially important class of mixed-moiety compounds is the huge family of glycol ethers, e. g., 2-ethoxy-ethanol, 2-(2-ethoxyethoxy)-ethanol, etc.

- Measurements on model mixtures over wide ranges of temperature and pressure – Many phase equilibria and thermophysical property measurements are restricted to a few properties over a limited range of T and P. Most useful to molecular simulators and the producers of estimation methods would be data on say, G^E, H^E, C_P^E, V^E, transport, and electromagnetic properties on a few carefully-chosen mixtures that represent the variety of intermolecular interactions encountered in "real" design problems. Another way to ask this question is: what should be the "steam-table equivalent" for mixtures? I suggest water + ethylene glycol.

- Data on amides – Thermophysical property data on pure amides are sparse, and mixture data are almost non-existent, probably because these are high-melting chemicals.

- Data for the UNIFAC binary interaction matrix – As discussed above, the amount of empty space in the UNIFAC matrix has been decreasing for the past 25 years. Perhaps a systematic examination of the mixture types that remain and a program to measure them is in order.

- Data measured at other than 25 °C – Examples of this problem are the many V^E measurement studies that are now reported, made easy by the advent of commercially avail-

---

[39] **Fredenslund, A., Gmehling, J., and P. Rasmussen (1977)**. *Vapor-Liquid Equilibria Using UNIFAC*. (New York: Elsevier), Table 4.4, pp. 43-47.

[40] **Nölker, K. and M. Roth (1998)**. *Modified UNIFAC Parameters for Mixtures with Isocyanates*. Chem. Eng. Sci., **53**(13), 2395-2401.

able vibrating-tube densimeters. Although measurements over the range 10 to 80 °C are very easily done with this instrument, most mixture V^E data are still only reported at 25 °C as if it were still being measured with pycnometers! The theoretically important temperature dependence and connection to S^E is not derivable from V^E data measured only at 25 °C.

- Liquid-phase heat capacity data – Measurement of liquid-phase heat capacity to ±1 to ±3 % for typical room-temperature liquid chemicals is straightforward with a differential scanning calorimeter (DSC). Yet few data by this method are reported. An important use of heat capacity data is thermodynamic extrapolation of vapor pressure data to lower temperatures.

- Environmental phase equilibrium data – Data for octanol-water partition coefficients, Henry's Law constants, and water solubilities (particularly as a function of temperature) are limited. More measurements would improve group-contribution estimation methods for these properties.

- Process safety data – Few systematically measured flash-point temperature and auto-ignition temperature data exist.

- Simultaneous physical and chemical equilibria data – This general topic would benefit from more attention by experimenters, simulators, and theoreticians: the analysis, measurement, and characterization of simultaneous chemical reaction equilibria and phase equilibria. This topic is rarely covered in engineering education, ignored in many experiments, and not available as an option in many process-simulator applications (or available as only an approximation). Jensen and Datta[41] point out, "It is, indeed, surprising that although the problem of determination of liquid-phase reaction equilibria from gas-phase thermodynamic data is common, it is not the stuff of textbooks yet." Work on this issue would help address the "nervous shuffling of feet" noted by Howard Hanley whenever chemistry was brought up during the Forum 2000 discussion.

---

[41] **Jensen, K. L. and R. Datta (1995)**. *Ethers from Ethanol. 1. Equilibrium Thermodynamic Analysis of the Liquid-Phase Ethyl tert-Butyl Ether Reaction (ETBE)*. Ind. Eng. Chem. Res., **34**(1), 392-399.

Is the Job Complete? Definitely not!

James C. Rainwater
Theory and Modeling of Fluids Group
National Institute of Standards and Technology
Boulder, Colorado 80305-3328, USA
James.Rainwater@Boulder.NIST.Gov

By "the job" in the title, I refer to the measurement of "conventional" thermodynamic properties of "simple, traditionally important" fluids. Speakers such as Bill Wakeham have pointed out opportunities for experimentalists to study untraditional substances and properties, and the community should be alert to such opportunities, but there is still much important work to be done in thermodynamic measurements as has been conducted over the last five or six decades. This essay is written from the point of view of a theorist who has searched for ways to adapt theoretical methods to problems in thermodynamics for which methods have traditionally been unavailable or not widely used.

By "conventional thermodynamic properties", I mean phase equilibria, PVT data, enthalpies, specific heats, and speeds of sound. Our group has a tradition of employing multi-parameter equations of state, such as the extended BWR equation, to characterize the complete thermodynamic surface of pure fluids and mixtures. Ideally, the input for such a correlation includes all of the above properties, which tend to complement each other in the information they contain. Such properties, of course, are also industrially important, some more than others.

To this list should be added, perhaps, transport properties, primarily shear viscosity and thermal conductivity, and secondarily diffusion coefficients and bulk viscosity. These properties are also of industrial importance, but are much more difficult to understand theoretically. I have been involved in a theory of the first density correction to the shear viscosity and thermal conductivity,[42] and at first found it surprising that such a theory, analogous to the second virial coefficient for pressure, was not widely available. This theory has generally met with success in describing experimental data,[43] but is not completely rigorous or definitive. For transport properties, the dilute gas and the critical region are well-understood, but for other density regimes there has been little support for development of theories. While theoretical approaches have not been exhausted, it seems that a rigorous theory, if developed, would involve integrals over all possible collisions over successively more molecules, and would require a computation comparable or greater in time and effort to a simulation.

The concept of a "simple and traditionally important fluid" is a bit harder to define, but, since fluids are best characterized initially by their critical temperature and pressure, I would propose an upper limit in critical temperature for such fluids of 600 K. It has been a tradition,

---

[42] **Rainwater, J. C. and D. G. Friend (1987).** *Second Viscosity and Thermal Conductivity Virial Coefficients of Gases: Extension to Low Reduced Temperatures.* Phys. Rev. A, **36**, 4062-4066.

[43] **Bich, E. and E. Vogel (1991).** *The Initial Density Dependence of Transport Properties: Noble Gases.* Int. J. Thermophys., **12**(1), 27-42.

unfortunately no longer followed worldwide as much as before, to measure experimentally the complete phase diagram of a fluid or a binary mixture, up to and including the critical point. Beyond 600 K, most organic molecules tend to decompose, which introduces a serious complication into thermophysical properties measurements.

I have been involved in a theoretical study of binary mixtures over an extended critical region, and within this project performed a comprehensive survey of VLE data as of 1991.[44] In that work I defined a "thoroughly measured mixture" as one for which there were at least four VLE isotherms or isopleths, measured up to the critical locus. I located 129 mixtures composed of 73 different pure fluids, and that list of fluids, along with some obvious omitted homologues, seemed to serve as a working list of simple and important fluids. While a few of the experiments were conducted at the end of the 19th century by the van der Waals laboratory, and in the early 20th century by Barnett F. Dodge of Yale, the beginning of an extensive worldwide effort took place in the 1940s with Webster B. Kay of Ohio State, Sage and Lacey of Caltech, and Donald Katz of Michigan, several students of whom developed their own productive laboratories. Shortly thereafter, large measurement programs were also initiated in London, Moscow, and elsewhere. Unfortunately, since 1994 this five-decade effort seems to have come to a halt, and I have not found a new thoroughly measured mixture in the literature since then, though clearly there are many important mixtures to be measured. New VLE laboratories have emerged in nations such as Korea, Taiwan, and China, but they have taken data over a less extensive temperature and pressure range.

The number of possible mixtures from even a short list of, say, 15 pure fluids has been pointed out by Bill Wakeham to be unmanageably large. This may be well-known, but is worth a formal explanation. How many mixtures can be formed from N pure fluids? To form the mixture, one either chooses or does not choose a fluid on the list, which leads to 2^N possibilities. One subtracts off the null result, i. e., no fluids chosen, and the N pures, in which just one is chosen, so the total number of mixtures is 2^N-N-1, and thus essentially grows exponentially with N. However, the hope is that binary interactions are predominant; it is known that true three-body interactions affect ternary mixtures, etc., but they yield a small contribution. Hopefully, an understanding of the thermodynamics of all important binary mixtures will lead to reliable predictions of multicomponent ones.

The thermophysical properties program of NIST/NBS-Boulder received substantial funding over a significant time interval for three major projects. Beginning in the 1950's cryogenic measurements were made for several decades on fluids such as liquid oxygen and liquid hydrogen important to the space program. This was followed by an extensive measurement program for petroleum-related fluids. In the late 1980's, prompted by worldwide concern for the ozone layer, NIST led the experimental effort to characterize alternative refrigerants. Substantial support was available out of significant national, economic, and environmental concerns, respectively.

---

44 **Rainwater, J. C. (1991)**, in *Supercritical Fluid Technology* (J. F. Ely and T. J. Bruno (eds.), Boca Raton, FL: CRC Press), p. 57.

Our group's apparatus has evolved accordingly. Until the late 1970's we were part of the Cryogenics Division, and the change from cryogenics was controversial at the time. But the limited number of substances that were fluid at cryogenic conditions had been thoroughly studied, and in retrospect it was appropriate to move on to substances with melting points higher than room temperature, and with apparatus designed for a higher temperature range.

However, since science evolves unpredictably, it was important that at least somewhere in the world some cryogenic apparatus was retained. Recently the theory of solid-liquid equilibria has achieved some maturity, but the initial choices of candidate systems are those of small molecules which generally freeze at cryogenic temperatures. A theory of SLE requires experimental information on the crystal structure, and it was noted that, until recently, the structure of frozen propane was unknown.[45] I was involved in a study of the freezing of methyl chloride.[46] Fortunately, just recently a group in Germany has determined the crystal structure of propane,[47] and one in Japan has reconfirmed the structure of methyl chloride in bulk and in small pores,[48] but these studies could not have been done if cryogenic apparatus had become extinct worldwide.

The situation is the same at higher temperatures. From our theory of the critical region of mixtures, it appeared that the critical locus of the important mixture carbon dioxide + propane was in error. That mixture was remeasured in our laboratory[49] and it was found that the literature values of critical pressures were in error by as much as 8 bars, an egregious discrepancy. A few years later, with the same theory we found evidence that, in a study of the phase diagram of ethylene + n-butane, the experimental compositions may have been substantially in error.[50] We recommended a remeasurement, but by that time the research and funding climate had changed, and a remeasurement was not undertaken.

Identification of scientifically needed or suspect data will be made with advances in theoretical methods, but will occur rather sporadically, and a larger, coherent project of properties measurement is desirable. In my view, the most glaring defect in available experimental data

---

[45] **Shen, W. N. and P. A. Monson (1995)**. *Solid-Fluid Equilibrium in a Nonlinear Hard Sphere Triatomic Model of Propane.* J. Chem. Phys., **103**(22), 9756-9762.

[46] **Gay, S. C., P. D. Beale, and J. C. Rainwater (1998)**. *Solid-Liquid Equilibrium of Dipolar Heteronuclear Hard Dumbbells in a Generalized van der Waals Theory: Application to Methyl Chloride.* J. Chem. Phys., **109**(16), 6820-6827.

[47] **Boese, R., H.-C. Weiss and D. Bläser (1999)**. *The Melting Point Alternation in the Short-Chain n-Alkanes: Single-Crystal X-Ray Analyses of Propane at 30 K and of n-Butane to n-Nonane at 90 K.* Angew. Chem. Int. Ed., **38**(7), 988-992.

[48] **Morishige, K. and K. Kanawa (1999)**. *Freezing and Melting of Methyl Chloride in a Single Cylindrical Pore: Anomalous Pore-Size Dependence of Phase-Transition Temperature.* J. Phys. Chem. B, **103**(37), 7906-7910.

[49] **Niesen, V. G. and J. C. Rainwater (1990)**. *Critical Locus, (Vapor + Liquid) Equilibria, and Coexisting Densities of (Carbon Dioxide + Propane) at Temperatures from 311 K to 361 K.* J. Chem. Thermodyn., **22**(8), 777-795.

[50] **Rainwater, J. C. and J. J. Lynch (1994)**. *A Nonlinear Correlation of High-Pressure Vapor-Liquid Equilibrium Data for Ethylene+n-Butane Showing Inconsistencies in Experimental Compositions.* Int. J. Thermophys., **15**(6), 1231-1239.

at present is thermodynamic properties (other than phase equilibria) for significantly dissimilar polar-nonpolar mixtures. By "significantly dissimilar," I mean that the critical points of the pure fluids, as well as their polarities, differ substantially. In the mixture packages constructed at NIST-Boulder, the package NIST14, as pointed out by Paul Mathias, basically includes only petroleum fluids which are mostly nonpolar, and nonpolar-nonpolar mixtures behave, for the most part, ideally in the sense that they obey Raoult's law. REFPROP includes traditional and alternative refrigerants of which some are nonpolar and some polar, but these all have roughly equivalent critical points and thus mix easily. Special mixing rules incorporating excess free energies, such as Huron-Vidal and Wong-Sandler, have been developed for phase equilibrium and cubic equations, but to characterize the entire mixture thermodynamic surface, they should be adapted to multi-parameter equations of state.

A mixture correlation should be verified by checking against all thermodynamic properties mentioned in the introduction for mixtures, but at present an extensive database is unavailable. A 1985 review of PVT data for pure liquids and mixtures[51] listed mostly nonpolar-nonpolar mixtures, with far fewer polar-polar mixtures and only a handful of polar-nonpolar mixtures, and there is little new data since then. A fair number of enthalpies of mixing of polar-nonpolar mixtures have been measured by the Brigham Young University group of J. Bevan Ott and colleagues and the University of Bristol group of Christopher Wormald, but there is a dearth of sound speed and other data for such mixtures. Even if one takes the position that simulation will inevitably replace experiment, the goal of predicting all mixture thermodynamics will clearly require real experimental data against which to validate the simulations, and particularly will require data for these most nonideal of mixtures.

I believe the most suitable candidate mixtures are those of nonpolar alkanes with polar oxygenated hydrocarbons such as alcohols, ketones, and ethers. Given the choices within these homologous series, we can systematically study effects of critical temperature difference and polarity difference. All thermodynamic properties listed in the introduction ideally should be measured. There are also measurement needs for the pure oxygenated hydrocarbons, but data on the mixtures would be the most valuable. This would provide a needed database to test theories of mixing rules for all properties and to verify simulation results. Other polar substances, for example the amines, might also be considered.

A very important and challenging topic is aqueous mixtures. Because of its comparably high critical pressure, 22.1 MPa, basically twice as high as any other common fluid, and its polarity, water forms mixtures with other common fluids which are extremely nonideal, and have been a huge challenge to correlate, but are of great interest. Alcohols, ketones, and ethers have the polarity but not the extremely high critical pressure of water, so correlations of mixtures of these oxygenated hydrocarbons with alkanes would serve as useful steppingstones to understanding mixtures of water and alkanes.

---

51   Tekac, V., I. Cibulka, and R. Holub (1985). *PVT Properties of Liquids and Liquid Mixtures: A Review of the Experimental Methods and the Literature Data.* Fluid Phase Equilib., **19**, 33-149.

Such an experimental program may seem less compelling to society, and thus not as readily funded, as the three long-term projects mentioned previously. But a large and reliable database of thermophysical properties of polar-nonpolar mixtures, in my view, is a needed prerequisite to a complete understanding of the thermodynamics of fluid mixtures and to realizing Peter Cummings' vision of the prediction of properties of fluid mixtures by reliable and verifiable simulations in the future.

Challenges in the Development of Transferable Force Fields for Phase Equilibrium Calculations

J. Ilja Siepmann
Departments of Chemistry and of Chemical Engineering and Materials Science
University of Minnesota
Minneapolis, Minnesota 55455-0431, USA
Siepmann@Chem.UMn.Edu

Over the next decade, it will become possible to gain microscopic-level insight into the behavior of complex chemical systems making use of theoretical advances and employing high-speed computational resources. Only with this molecular-based understanding will the research enterprise be able to develop chemicals, materials and processes that meet the increasing needs of society. The "computer experiment" is unique in the sense that it enables the study of well-defined systems under well-controlled physical conditions with a non-invasive approach. The simulator can specify as input parameters the molecular structure of the constituents, their concentrations, pressure and temperature, and then follow the phase space trajectory of the system. Analysis of the trajectory allows us to determine mechanical and thermal properties and ultimately to learn about how molecular architecture and composition influence function.

Knowledge of phase equilibria and thermophysical properties are of prime importance for all areas of chemical engineering and chemistry, and it is therefore not surprising that the experimental determination and modeling of their behavior has been paramount to research in the chemical sciences. Although the thermophysical properties of pure low-molecular-weight hydrocarbons have in general been determined experimentally, many important areas remain difficult and costly to access by experimentation. Experimental difficulties are caused by the necessity to study systems at high temperature and pressure, the lack of pure samples, and the immense variety of technologically important fluid mixtures. Molecular modeling and other theoretical prediction methods are therefore needed to complement the available experimental data and to improve the rational design of efficient chemical processes.

Predicting phase equilibria and other thermophysical properties of multicomponent mixtures, given only the architecture of the molecules (types of atoms and their connectivity) and the experimental conditions, remains one of the grand challenges for the field of molecular simulation. The success of molecular simulation in this endeavor depends on the availability of efficient simulation algorithms and accurate force fields. For a long time, progress was limited by the available simulation techniques and by computer processing speed. Over the last 15 years, however, enormous advances have been made in simulation methods, and it is now becoming evident that the development of sufficiently accurate force fields deserves more attention. Already in 1982, Rowlinson and Swinton wrote: "Both experiment and statistical theory have reached so high a degree of accuracy and sophistication that they put too great a strain on the weak link between them—our knowledge of intermolecular energies. It is improvement in this area that is needed next."[52]

The remainder of this essay is devoted to the author's opinions on the following questions concerning the development of transferable force fields: (i) whether these force fields should sacrifice simplicity for improved accuracy; (ii) whether these force fields should use "true" potentials or "effective" potentials; and (iii) whether these force fields should be developed by empirical means (using comparisons with experimental data) or starting from electronic structure calculations. The focus of this discussion is solely on "non-bonded" interactions, because most improvements in the development of transferable force fields are made in this area.

As a starting point, it should be stressed that the term "transferable" implies that the force field parameters for a given interaction site should be transferable between different molecules (e.g., identical parameters should be used for the methyl group in, say, *n*-hexane, 1-hexene, or 1-hexanol) and that the force field should be transferable to different state points (e.g., pressure, temperature, or composition) and to different properties (e.g., thermodynamic, structural, or transport). Thus, while a force field that uses special types of interaction parameters for specific molecules or special combining rules for specific unlike interactions, might be very accurate for specific applications, it would not be considered transferable and would most likely have limited predictive power for unrelated applications.

In the development of force fields, one has to achieve a balance between competing virtues: on the one hand, there should be a quest for simplicity and computational efficiency (regarding numbers and types of interaction sites, functional form of interaction potential, number of adjustable parameters). On the other hand, one is in pursuit of accuracy and transferability. While simplicity and accuracy are competing virtues, however, they are not mutually exclusive. That is, a judicious choice of force field parameters and potential functions allows one to improve accuracy without sacrificing simplicity. This is most evident in recent developments of united-atom force fields for hydrocarbons, which have lead to significant gains in accuracy without sacrificing the simplicity of the original united-atom model of Ryckaert and Bellemans.

The question on the use of "true" or "effective" potentials is directly connected to the chosen level of simplicity of a force field. For example, if one decides to use simple pairwise additive potentials for simulations of condensed phases, then the best strategy is to derive parameters for an "effective" potential (e.g., the dispersive component of the pair potential should include many-body corrections and the charge distribution should reflect induced polarization). One should never ignore the fact that the Lennard-Jones potential is only a poor approximation of the true van der Waals interactions and that placing partial charges on atomic sites is only a poor representation of the true molecular charge distribution. Thus increasing the number of interaction sites and accounting for many body dispersion and induced polarization allows one to move toward the "true" potential, but it should be recognized that the overwhelming majority of molecular mechanics force fields is based on "effective" potentials. Regarding the question whether transferable force fields should be developed by empirical means or based

---

[52] **Rowlinson, J. S., and F. L. Swinton (1982).** *Liquids and Liquid Mixtures* (London: Butterworth Scientific).

on electronic structure calculations, the answer again depends on the simplicity of the force field. Fitting to available experimental data appears to be the better strategy for a simple force field with "effective" potentials. However, the use of high-level electronic structure calculations is clearly the way to go for the most accurate force fields with the best possible description of the "true" interactions. Even for the simple force fields, however, information from electronic structure calculations should be utilized to narrow the parameter space for the empirical fitting procedure. Finally, considering the wide range of required accuracies for different applications and the different complexities of the chemical system to be investigated, it becomes obvious that a single force field cannot satisfy all demands. Thus it is advantageous to create a family of force fields that can cover a spectrum of accuracy requirements and system complexities.

A Perspective on Connecting Virtual Design with Physical Reality

Lambert J. Van Poolen[53]
Engineering Department
Calvin College
Grand Rapids, Michigan 49546, USA
`VPol@Calvin.Edu`

We start with a real story to set the stage for our perspective on how it is that virtual design (computer simulation) connects with physical reality (performance of real machines). Even though the primary focus of Forum 2000 was on the connection between computer modeling of fluids and experimental data, ultimately the issue is how it is we can model reality and use resultant computer simulations to predict the performance of real machines.

The heart of our story lies in the plastic injection-molding business. Injection-molded plastic parts are important components, for example, in automotive interior systems. The ability to predict how the combination of machine, tool, and material will perform in the manufacturing process of plastic parts is important. For example, to be able to predict the filling and packing pressures to achieve proper part density is important for ending up, efficiently and economically, with the intended part dimensions. As plastic parts shrink to a solid upon cooling, their initial liquid-melt density (after packing) determines the volumetric shrinkage that, in turn, determines directional, linear shrinkage. With the increasing number of different, complex parts and an emphasis on newer, less expensive materials, experience alone was inadequate to predict fill and pack pressures.

To address this issue, a group of engineers and scientists at a major research university obtained a sizable National Science Foundation grant to, in essence, develop a model of the plastic injection molding machine. This (finite-element) model included the fluid-flow and heat-transfer characteristics of plastics. Various thermodynamic and transport properties needed for these models were measured in the laboratory and fitted to essentially empirical equations. Examples of properties measured and represented in various computational forms are the pressure-volume-temperature (PVT) data for the liquid and solid plastic and viscosity data for the liquid plastic. With models and data in hand, senior scientists and engineers left the university and started their own company. Software to simulate the plastic injection-molding process was sold worldwide. More importantly for our discussion, their visionary goal was to predict the behavior of real combinations of part, machine, and material with just the theoretical, universal model *including related laboratory data*. The ideal was that no allowance need be made for individual peculiarities in the wide variety of particular combinations of part, machine, and material. One universal model was to fit all particular situations. The end of the story is that this goal was never attained. The business was sold to another group whose best-selling simulation software is calibrated to match the performance of actual molding

---

[53] Also member of the Advanced Engineering Group, Johnson Controls (Automotive Interiors) Inc., Holland, Michigan, and guest scientist in the Physical and Chemical Properties Division, National Institute of Standards and Technology, Boulder, Colorado.

machines.

What lesson is learned? Although the above is just one story, it does comport with our experiences in engineering design: models (including laboratory data) cannot be used alone to predict performance of real machines. For insights into why this is so we turn to the work of the philosopher, scientist, engineer, and McArthur fellow, Nancy Cartwright.[54] Her basic contention is that the laws of science *only* apply to *models* of reality. To obtain computational, predictive forms or equations, we must first develop physical models that transform airplane bodies into simple cylinders or molecular gas interactions into simple potentials. The physical laws (Newton's, etc.) are *then applied to these models* to develop computational forms to estimate, for example, stresses in the aircraft fuselage or the pressure-volume-temperature (PVT) relationship for a gas. A reduction (of reality) of sorts takes place necessarily in order to end up with computational forms. A reduced reality must be utilized to eventually obtain equations that allow some measure of estimation for the size, shape, and material of machines in the making. In her book[55] she forcefully makes her point: "…theories in physics do not generally represent what happens in the world; only models represent in this way …" The result of this necessary, reductive procedure is that the computational equations stand at some distance from real, particular situations—the actual airplane or gas. The resultant equations take on a universal appeal that, paraphrasing Cartwright, "abstract from all the particular situations to provide a general description common to all, but is not literally true of any of them."

Our practice of engineering design leads us to agree. At best, we use universal, textbook models (and equations) as rough and ready guidelines for design, but cannot use them to finally determine the size, shape, and material of machines. Practically speaking, we dare not assume that the results of computer simulations using these models give us the actual answers to our design questions. As a result, in practice, we often calibrate these models with information germane to our particular situation. This increases our confidence in the use of models especially as guides for design decisions. In more philosophical language, we suggest that universal theoretical models require calibration for use in predicting the performance of particular, real machines.

An example of the interplay between model and calibration is the prediction of the linear shrinkage of plastic parts upon cooling to a solid. For isotropic behavior, this shrinkage is 1/3 the volumetric shrinkage predicted from PVT (laboratory data-based) equations. Actual dimensional measurements however, indicate that, for the particular resins we use, the number instead should lie in a range from 1/9 to 1/12. The actual combination of part geometry and material does not shrink isotropically. We then modify predictions obtained from the models with particularized information to predict the real behavior of similar parts. For sun-visor frames, made of polypropylene, there is one calibration constant, for door panels, made of ABS, we need a different calibration constant. Theoretical, universal models are not eschewed. However, they need to be modified (particularized) if they are to be used to confidently predict real machine

---

[54] **N. Cartwright (1999)**. *The Dappled World: A Study of the Boundaries of Science*, Cambridge University Press.

[55] *Ibid*, p. 180.

performance. And, as Cartwright suggests, this is *necessary* because models cannot, by definition, match reality. The model is a carefully constructed *reduction* of physical reality (of necessity), and we apply scientific laws to it, to obtain forms useful for calculation. The results must be used intelligently by engineering designers if their machines are to function properly in reality.

As I understand it, a basic question within Forum 2000 was could we simulate, via the computer, the real world? And further, can we do that without experimental data? The author's answer is an emphatic "No!" to both questions. Any theoretical model is already quite an abstraction from reality. At minimum, these models must be validated within the assumptions made in their development. This means that they must be tried out in the laboratory under the conditions assumed. (How much validation data are needed is another question and is not addressed here.) Certainly, what level of confidence a design engineer places in models is directly related to laboratory data and theory going hand in hand to produce computational forms.

Yet, the engineer must often modify the computational forms (based on theory and concomitant laboratory data) in order to use them in design. One of the reasons for this is that most often, an experimental apparatus tends to be itself *designed to fit the theoretical model*. A case in point is the laboratory measurement of the viscosity of plastic melt. It is done in a standardized apparatus—a simple flow tube that looks a lot like what one would assume for a physical flow model. Data obtained *conform to that model*. These experimental data then fit the description of models by Cartwright above: "They relate to all situations but apply to no one in particular." In real molding machines, the plastic can flow quite differently from how it does in a simple tube. As a result, we have developed a method of measuring plastic melt viscosity for particular machine/tool/resin situations utilizing information from actual machines in operation. We then combine these particularized viscosities with the universal flow models (in the simulation software) to better predict filling pressure requirements.

Such changes to the models are necessary to align them with reality. Often large sums of money are at stake as well as, in many cases, the safety and welfare of people. Hence, we need to make *real predictions*. Theoretical models coupled with experimental laboratory data were not adequate to predict the performance of complex, individual machines. Certainly, experimental measurements bring theoretical models closer to reality; however, while important, that was not enough. Calibration of these validated models is needed for their use in engineering design decisions.

I have used both the experiences with plastic injection molding and the ideas of Nancy Cartwright to argue that the use of theoretical models (and their embodiment in computer simulations) cannot be used alone to predict the performance of particular machines. The argument really is that, by definition, the very nature of the model is such that it is necessarily several steps removed from reality. Experimental data help move abstract, theoretical models closer to matching performances of real size, shape, and material combinations that we call machines. Validated models can and are used for preliminary design predictions. Yet, when precise information is needed about a particular design, the gap between model and reality (real machine performance) can be effectively closed by means of efficient reality-matching calibration procedures of relatively low cost.

In summary, it should be clear from an engineering (and philosophical) point of view that predictive tools for design decisions rely on theoretical models validated in the laboratory. And where necessary, these are further modified by calibration to match the performance of particular machines. At best, in my experience, we can hope only for combinations of model, data, and calibration that provide predictive capacity for sub-classes of these many combinations. Of necessity, there can be no universal simulation model (independent of data) to predict performance of real machines. We must supply connections between our theoretical constructs (necessary reductions of reality) and the real machines to be designed. Experimental data and calibration are essential elements in that connection between virtual design and physical reality.

www.ingramcontent.com/pod-product-compliance
Lightning Source LLC
Chambersburg PA
CBHW080259180526
45167CB00006B/2588